A First Guide to Computational Modelling in Physics

This innovative text helps demystify numerical modelling for early-stage physics and engineering students. It takes a hands-on, project-based approach, with each project focusing on an intriguing physics problem taken from classical mechanics, electrodynamics, thermodynamics, astrophysics, and quantum mechanics. To solve these problems, students must apply different numerical methods for themselves, building up their knowledge and practical skills organically. Each project includes a discussion of the fundamentals, the mathematical formulation of the problem, an introduction to the numerical methods and algorithms, and exercises, with solutions available to instructors. The methods presented focus primarily on differential equations, both ordinary and partial, as well as basic mathematical operations. Developed over many years of teaching a computational modelling course, this stand-alone book equips students with an essential numerical modelling toolkit for today's data-driven landscape, and gives them new ways to explore science and engineering.

PAWEŁ SCHAROCH is University Professor within the Department of Semiconductor Materials Engineering at Wrocław University of Science and Technology, Poland. He has also held positions at Durham University, the Fritz Haber Institute, Berlin, and at Orange Labs Poland. His principal work is on structural and electronic properties of atomic systems. Professor Scharoch also has around thirty years of experience in teaching various courses in general physics and computational physics.

MACIEJ P. POLAK holds a PhD in Physics from Wrocław University of Science and Technology and is currently a dedicated researcher at University of Wisconsin–Madison's Department of Materials Science and Engineering. He works on first-principles modelling of the electronic band structure of highly mismatched semiconductor alloys for their use in opto-electronics and metals for space applications devices. With over thirty published peer-reviewed articles, he consistently strives to push the boundaries in scientific exploration.

RADOSŁAW SZYMON, MSc, graduated with honours from Wrocław University of Science and Technology in 2022. He is currently pursuing a PhD in semiconductor technology at the same institution, supported by the prestigious Pearl of Science grant. He also enjoys conducting physics simulations, in particular in electromagnetism and quantum mechanics.

A First Guide to Computational Modelling in Physics

Paweł Scharoch
Wrocław University of
Science and Technology

Maciej P. Polak
University of
Wisconsin–Madison

Radosław Szymon
Wrocław University of
Science and Technology

Software developed by

Katarzyna Hołodnik-Małecka
Wrocław University of
Science and Technology

Shaftesbury Road, Cambridge CB2 8EA, United Kingdom

One Liberty Plaza, 20th Floor, New York, NY 10006, USA

477 Williamstown Road, Port Melbourne, VIC 3207, Australia

314–321, 3rd Floor, Plot 3, Splendor Forum, Jasola District Centre,
New Delhi – 110025, India

103 Penang Road, #05–06/07, Visioncrest Commercial, Singapore 238467

Cambridge University Press is part of Cambridge University Press & Assessment,
a department of the University of Cambridge.

We share the University's mission to contribute to society through the pursuit of
education, learning and research at the highest international levels of excellence.

www.cambridge.org
Information on this title: www.cambridge.org/9781009413121

DOI: 10.1017/9781009413138

First published 2024

A catalogue record for this publication is available from the British Library

*A Cataloging-in-Publication data record for this book is available from the Library of
Congress*

ISBN 978-1-009-41312-1 Hardback
ISBN 978-1-009-41310-7 Paperback

Cambridge University Press & Assessment has no responsibility for the persistence
or accuracy of URLs for external or third-party internet websites referred to in this
publication and does not guarantee that any content on such websites is, or will
remain, accurate or appropriate.

Contents

Preface

Numerical modelling is a relatively recent and powerful scientific tool that has experienced remarkable development since the mid-twentieth century, fuelled by advancements in computer technology. Today, it is difficult to find a field in science or engineering where computational methods do not play a crucial role. The list of benefits is extensive, including new opportunities such as predicting the properties of physical systems, exploring properties that are not experimentally accessible, obtaining quantitative data that require massive amounts of mathematical operations, processing vast amounts of data, and facilitating machine learning, among others. These new opportunities are accompanied by the relative ease of application, rapid acquisition of valuable results, and cost-effectiveness. As a result of these features, computational methods have become an independent scientific tool, complementing experiment and theory. On the one hand, they extend mathematical modelling and could not exist without it; on the other hand, they establish their own methodology, often resembling experimental work through the use of virtual systems. Computational methods have proven to be invaluable in supporting experimental research, technological advancements, and even theoretical physics, where mathematical models often require extensive numerical calculations to yield results in the form of quantitative data. Considering the aforementioned points, it is clear that various aspects of numerical modelling should be an integral part of academic education, particularly in the fields of science and engineering. In fact, universities worldwide have been offering such educational components for years, often as independent specialisations, such as Applied Computer Science or Big Data. These specialisations focus on diverse aspects of computer applications in science and technology, with an emphasis on numerical modelling. This course is designed to address the need for numerical modelling education. Its level is tailored so that first-year physics or engineering students can fully engage with the material. It is crucial to introduce these methods early in higher education, as they continue to grow in popularity. The course employs a project-oriented teaching approach, as suggested by the title *A First Guide to Computational Modelling in Physics*. Rather than systematically covering numerical methods, the course introduces them as tools necessary for solving specific problems. This approach makes learning the methods more engaging and goal-oriented. Aside from the 'First Steps' section, which teaches students numerical methods for basic mathematical operations (finding zeros and extrema of a function), the course is divided into eight basic and six advanced projects. Each basic project begins with an introduction to the necessary physics background, followed by a presentation and discussion of a specific problem and its mathematical description. Next, the required numerical methods and algorithms are introduced, and finally, a set of exercises is proposed. Advanced projects build upon the previously introduced physics background and provide opportunities for students to conduct their first computational investigations, allowing for the realisation of individual research scenarios. This book is designed as a stand-alone resource, containing all the necessary materials for the completion of each project. While there is no strict requirement to use other sources, students are strongly encouraged to expand their knowledge base by consulting additional literature, as this is an inherent aspect of higher education.

The methods presented in this book primarily focus on differential equations, both ordinary (ODE) and partial (PDE), while also covering basic mathematical operations, derivatives, and quadrature, as introduced in the second project (Diffraction on a Slit). The book addresses typical problems such as initial value problems (IVP), boundary value problems (BVP), and eigenvalue problems (EVP). Most of the methods discussed are based on variable discretisation and recursive algorithms for ODEs. For PDEs, the methods of finite difference (FD) and finite elements (FE) are explained, along with selected matrix methods for solving systems of equations (FD) and chosen iterative optimisation methods (FE). In the case of PDEs, system symmetry is utilised to reduce dimensionality from 3D to 1D, simplifying implementation and allowing projects to be completed within reasonable time constraints.

An additional benefit of this course is the effective learning of physics, as it offers strong motivation for revisiting and expanding upon one's knowledge. Each project is prefaced by a brief presentation of the relevant physical background. While these presentations are not exhaustive, they are designed to be coherent, provide sufficient information for project execution, and inspire deeper study.

The authors would like to extend their gratitude to Anna Latosinska for her creative review of the and to Dawid Dworzanski for developing the project 'Evolution of a Wave Function in a 1D Quantum Well', and to Kamil Wrzos and Piotr Tokarczyk for their contribution to the project 'Hydrogen Star'. Additionally, they acknowledge the significant contributions of Technical Physics and Quantum Engineering students from the Faculty of Fundamental Problems of Technology at Wroclaw University of Science and Technology, who contributed to development and verification of computer codes in C++ and Python.

How to Use the Book

Students participating in this course are expected to possess a general understanding of physics at an academic level, be familiar with a chosen operating system, know how to use a selected graphics application, and have basic programming skills in a chosen language.

Before embarking on the projects, it is recommended that students read the 'First Steps' chapter and complete the exercises therein. The basic mathematical operations discussed in this chapter (finding zeros and extrema of a function) are relatively simple yet highly useful. Completing these exercises will familiarise students with the working environment, including the computer, operating system, programming environment, and graphical tools, thereby making it easier to work on subsequent projects. The first eight projects are *Basic*, each containing four sections: 1. Physical Background, 2. The Problem, 3. Numerical Methods, and 4. Exercises. Project 4 includes an additional section, 'Reduction of a Single Planet Motion in a Central Field to 1D'. Advanced projects (9–14) build upon the physical background introduced earlier and typically contain only problem definitions, occasionally featuring a description of the numerical algorithm if not previously introduced, and exercises. When working on basic projects, students should begin by reviewing or learning the physics background and familiarising themselves with the physical system under consideration, along with its mathematical description (Sections 1 and 2). If necessary, students should consult additional sources to address any difficulties in understanding, although the material provided in the book should be sufficient for completing the projects. The next section, 'Numerical Methods', is crucial to achieving the primary goal of the course. In this section, numerical algorithms are either fully derived or their underlying concepts are explained to facilitate their conscientious use and help prevent errors arising from the digital character of analysis. Students are also encouraged to derive numerical formulas on their own, following the given ideas (which often appear as exercises). Understanding the rationale behind a specific numerical formula is more important than being able to derive it. The final section contains exercises that should be completed using a computer program.

Exercises in this course are divided into three categories: *basic* (which should be completed by all students), *supplementary* (designed to deepen and strengthen knowledge and skills), and *advanced* (intended for students with a particular interest in the subject matter). Within the basic exercises, emphasis is placed on crucial steps in computational work:

1. **Testing the program**. This involves running the program for cases where the results are known from other sources (e.g., analytical solutions) and verifying that the program reproduces these results. This step is typically associated with the initial runs of the program after the code has been written and cleared of any errors that may have arisen.

2. **Testing the effect of control parameters and establishing their correct values**. This step is an essential aspect of computational work. The creation and implementation of numerical algorithms always involve the digitisation of analog formulas, which introduces technical parameters controlling the procedure, such as time steps or grid parameters for spatial variables. Incorrect values for these parameters can adversely affect the results and lead to errors. A fundamental method for assessing the impact of control parameters is the convergence test, which examines results as a function of a given parameter value. The issue becomes more complex when multiple control parameters are involved, as correlations between them may exist. However, the aim of this course is mostly to highlight the problem. Methods for evaluating the influence of control parameters by examining physical quantities (e.g., conservation laws) will be further described in the book.

3. **Virtual experiments**. This stage represents the primary objective of all preceding work. Once we are confident that the program is functioning correctly and have determined the appropriate values for control parameters, we can begin investigating the properties of a system. This process usually involves conducting experiments on a virtual system, observing its behaviour under various physical conditions (e.g. the motion of planets in a planetary system with different initial conditions).

Preliminary versions of the codes can be found in the online repository.

Figure 0.1 Internet repository; *https://wppt.pwr.edu.pl/ PhysModelCodes*

It is recommended that students write the first two to three codes in their chosen programming language to test and practice their programming skills and, if necessary, expand their knowledge. Model codes have been prepared in Python, complete with detailed descriptions (Python Notebooks), but Fortran and C++ versions are also available if preferred. The programs have been designed for clarity rather than optimisation in terms of efficiency. Some exercises require code modifications, providing an opportunity to practice programming skills further.

In addition to the eight basic projects, the book offers six advanced projects. These projects allow students to apply the methods learned throughout the course to the analysis of more complex physical systems. It is recommended that students with a particular interest in computational research undertake a chosen advanced project individually. All work throughout the course can be completed using free tools available on the Internet (e.g. `Spyder` for Python, `Force` for FORTRAN, `Gnuplot` for graphics).

First Steps

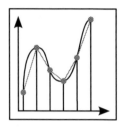

During this first computer lab session, students will learn or review the basic elements of their chosen programming language, programming environment, and graphical program. They will be introduced to numerical methods for fundamental mathematical operations, such as finding the roots and the minimum or maximum of a 1D function. Furthermore, students will practice essential programming operations, including loop instructions and tabulating data from 1D and 2D functions.

Basic Mathematical Operations

0.1 Finding Roots of a 1D Function

0.1.1 Bisection Method

Let's start from the case where we are sure that a single root of a function $f(x)$ is somewhere within the interval (x_l, x_r) (Figure 0.2). That means we already know the root of the function with an uncertainty of $\Delta x = (x_r - x_l)$. To reduce the uncertainty, we cut the interval from the previous step in half, and find the point in the middle $x_m = (x_r - x_l)/2$. We then get two new intervals, two times shorter than the previous one. However, the root of the function may be present only in one of them (it may happen that x_m is already the root, and this should be checked in the program). In order to identify the one within which the function crosses zero we check the condition $f(x_m) \cdot f(x_r) < 0$ (the function changes its sign). If the condition is true, the root is in the (x_l, x_r) interval, if not, then it is in the other one. We denote the ends of the new interval again by (x_l, x_r) by replacing $x_m = x_l$ (or $x_m = x_r$) and repeat the procedure. After n steps, the length of the domain segment containing zero is reduced by of factor 2^n and when it is less than the assumed

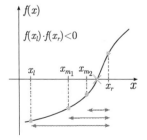

Figure 0.2 The bisection method

uncertainty ϵ, that is the condition $\Delta x < \epsilon$ is fulfilled, the whole procedure stops. Alternatively, we can predict how many steps (n) are needed
from the condition that after n steps the initial interval is reduced by a
factor of 2^n, and perform only this number of steps.

0.1.2 Secant and Newton–Rhapson Methods

The bisection method is very safe but not the most efficient. The
popular (and often more effective) alternatives are the secant and
Newton–Raphson methods (Figure 0.3). The Newton–Raphson method
is useful when together with the function $f(x)$ its analytical derivative is
also known. Using the derivative we construct a linear approximation
of the function $f(x)$ at the point x_n. Then we approximate the root of
$f(x)$ by the zero of the linear function (x_{n+1}), $x_{n+1} = x_n - f(x_n)/f'(x_n)$.
This root is treated as the new starting point and the iteration procedure
continues until a few subsequent changes in x_n are smaller than the
assumed uncertainty ϵ. The algorithm can be applied only if a power
series expansion of the function $f(x)$ in the vicinity of its root contains
a linear term. Of course, we usually do not know in advance if this is
the case. Thus this procedure, although quickly convergent, should be
applied with caution. The secant method is similar to Newton–Raphson
except that the derivative of a function is found numerically, using, for
example a three-point scheme (see Project 2). However, it should be
noted that this scheme requires several evaluations of the function $f(x)$
at each iteration step, and this may appear to be less efficient than the
safer bisection method.

Figure 0.3 The Newton–
Raphson method

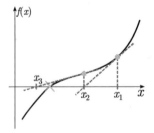

0.2 Finding Minimum of a 1D Function

0.2.1 Golden Section Search

Determining the minimum (or maximum) of a 1D function is a crucial
task, as many optimisation methods in multidimensional spaces rely on
directional minimisation. It is worth noting that finding the minimum

is equivalent to finding the maximum since changing the sign of the function converts maxima to minima and vice versa. Therefore, our discussion will primarily focus on identifying the minimum.

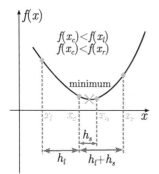

Figure 0.4 The Golden Section Search

To identify the domain interval containing a minimum in the interval, assuming that there is only one, we need a checking point x_c within the interval (x_l, x_r) (Figure 0.4). We can be sure that the minimum is present in the interval if $f(x_c) < f(x_l)$ and $f(x_c) < f(x_r)$. In constructing the algorithm one should focus on a proper choice of the new, fourth checking point x_n. By introducing the fourth point in the interval (x_l, x_r) we obtain two overlapping regions, each defined by three points. We can then identify the one containing the minimum using the condition above. To assure the optimal convergence, the two regions should have the shortest possible and equal length (at each step). This can be achieved if at each step the fourth point splits the longer of the two segments (x_l, x_c) and (x_c, x_r) in golden section ratio, that is $h_l/(h_l + h_s) = h_s/h_l$ (h_l is the longer segment, h_s is the shorter segment). By substituting $x = h_s/h_l$ into the above golden section condition we get the equation $x^2 + x - 1 = 0$, whose one of the solutions $(\sqrt{5} - 1)/2 \approx 0.62$ is the Golden Section proportion, known in art and architecture from ancient times. Thus, after n steps the length of starting interval reduces by factor $(0.62)^n$, which is a little bit slower than in the bisection algorithm of finding the root. The Golden Section Search is an always convergent, safe, and effective method of finding a minimum, but it is not the most efficient.

0.2.2 Other Methods

If the analytical form of the first derivative is known, then the minimum (or maximum) can be found by finding a root of the derivative. The derivative can be also approximated by a numerical formula.

The method of parabolas is based on a parabolic approximation, is analogous to the secant method for finding roots, and is often very effective. Using three points, x_l, x_c, x_r we can unambiguously locally approximate the function with a parabola. The position of the parabola's minimum is already an approximation of the function minimum but can be treated as the fourth control point x_c. The condition $f(x_c) < f(x_l)$ and $f(x_c) < f(x_r)$ allows to identify the segment containing the minimum. The function is then approximated with a parabola again in the new interval, and the procedure is repeated until a required uncertainty ϵ, in relation to the length of the last interval, is achieved.

Figure 0.5 The Simplex method

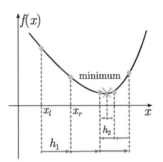

The 1D Simplex method (Figure 0.5) is the simplest method to search for the minimum which uses a *test window* rather than three points. Starting from a certain point, the domain is scanned in the direction where function decreases, let's say to the right, by moving a window (x_l, x_r) of certain length $h = (x_r - x_l)$. The domain interval containing minimum is found if the function begins to increase, that is $f(x_l) < f(x_r)$. At this point the window length is cut in half and the search continues from the x_r where the increase has been noticed, but in the opposite direction, until the condition $f(x_r) < f(x_l)$ is fulfilled, which, again, is a signal to turn around. The procedure continues until $h < \epsilon$, where ϵ is the assumed uncertainty. A great advantage of the method is that it can serve for finding interval containing the minimum in general, and then other, more effective methods can be applied to find it precisely.

0.3 Exercises

Obligatory

1. Using the program FTABLE tabulate your own function and visualise it with a graphical program.

2. Modify the FTABLE code so that it can tabulate a 2D function; set your own function and visualise it.
3. Test the BISEC code by finding the roots of a chosen second-order polynomial and comparing the results with analytical solutions.
4. Find the value of the number π as a zero of the $\sin(x)$ function. What precision can you achieve? Explain why.

Advanced

1. Write a 1DMINIMUM code that finds the minimum of a 1D function using one of the algorithms presented above (Golden Search, Parabolas or 1D Simplex). Test the program by finding the minimum of a chosen second-order polynomial and compare the results with analytical solution. Find the value of number π as a minimum of $\cos(x)$ function. What precision can you achieve? Explain why.

Project 1

Rectangular Finite Quantum Well – Stationary Schrödinger Equation in 1D

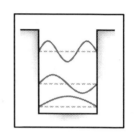

In this project, participants utilise the procedure of finding roots of functions to solve the eigenvalue problem of a rectangular quantum well (QWELL code). When considering the rectangular quantum well as the simplest model of a hydrogen atom, the code can be applied to determine its first two to three energy levels, which is the primary exercise in the project. The eigenvalue problem itself, appearing in various areas of physics (such as vibration mechanics, wave optics, and quantum mechanics), will be the subject of a separate project (Project 6) and one of the advanced projects (Project 12), where a rectangular quantum well partially filled with electrons is examined. It is somewhat paradoxical that despite employing the simplest mathematical operations in the current project, it is based on advanced physical concepts, such as quantum mechanics, often unfamiliar to first-year students. Learning the basics of quantum mechanics typically requires a 30-hour course and knowledge of advanced mathematics. Therefore, we will only introduce its fundamental and most straightforward ideas here, just enough to enable the conscious execution of the project.

1.1 Physics Background: Chosen Ideas of Quantum Mechanics

In quantum mechanics, while the physical quantities of interest, such as position or momentum of a particle, remain the same as in classical physics, their representation is entirely different. Focusing on the problem of a single particle, the central object is the quantum state rather than the coordinates in the chosen system (as it would be in classical physics). In the so-called position representation, the quantum state is a particular function of the position variable $\psi(r)$, which, from a mathematical perspective, must meet special conditions of differentiability and integrability. The function itself does not have a physical interpretation, but its squared modulus $|\psi|^2$ does – it represents the probability density of finding the particle at a given point in space, that is becomes a probability when multiplied by the volume element (the Born probabilistic interpretation). Here lies the main difference between classical and quantum physics – the probabilistic nature of the latter, with the concept of probability being inherent to the theory. When measuring a physical quantity, the outcome can only be predicted with a certain probability. The deterministic nature of phenomena, justified in classical physics, no longer holds, and this fact was challenging for many physicists to accept during the early stages of quantum theory development. For instance, Albert Einstein proposed the hidden variables hypothesis, suggesting that there are unknown variables that determine the measurement results. Modern interpretations, such as the Copenhagen Interpretation, go even further, positing that a particle can simultaneously exist in multiple positions with different probabilities (which is entirely impossible in the classical world), and only the act of measurement localises it to a specific position (e.g. the role of the measurement instrument is played by the screen in the 'electron diffraction on a double slit' experiment). The same concept applies to other physical quantities, meaning that quantum systems can simultaneously exist in various states of a particular quantity (with different probabilities), and during the measurement, the system selects one of these states. Currently, quantum mechanics is a coherent and complete theory, with the Copenhagen Interpretation being widely accepted, and no scientific evidence has emerged to challenge its validity.

The central and historically first equation for evaluating the state function is the Schrödinger equation

$$\left[-\frac{\hbar^2}{2m}\nabla^2 + V(r)\right]\psi(r, t) = i\hbar\frac{\partial\psi(r, t)}{\partial t}, \tag{1.1.1}$$

where $\hbar = h/2\pi$, h is Planck's constant, m mass of the particle, $\nabla^2 = \frac{\partial^2}{\partial x^2} + \frac{\partial^2}{\partial y^2} + \frac{\partial^2}{\partial z^2}$ is Laplace's operator, and $V(r)$ the particle potential energy.

The equation resembles a wave equation, which is why the function $\psi(r)$ is also called the wave function. As we can see, this is a function of both space and time variables. However, when the left-hand side of the equation (potential energy) does not explicitly depend on time, the function can be represented as a product of a space variable and time-dependent parts, with the latter having a known form $\psi(r)e^{i(\omega t)}$ (using the Euler representation of complex numbers, see Appendix A.1). A similar situation has been described (with respective derivation) in Project 6 for the case of a standing wave. If we substitute the factorised function into Eq. 1.1.1, we can easily eliminate the time-dependent part, which leads to the stationary Schrödinger equation

$$\left[-\frac{\hbar^2}{2m}\nabla^2 + V(r) \right] \psi(r) = \varepsilon\psi(r), \qquad (1.1.2)$$

where $\varepsilon = \hbar\omega$.

This equation is a starting point of the project. From a mathematical point of view it is an eigenvalue problem, the solution to which is a set of pairs: eigenvalues and corresponding eigenfunctions obeying the imposed boundary conditions, $\{(\varepsilon_n, \psi_n(r))\}$. The solutions are indexed with the integer n called the quantum number. The operator appearing on the left-hand side represents the total energy of the particle (Hamiltonian), and the eigenvalue problem leads to eigenenergies and eigenfunctions of the particle. From this an interpretation follows – the quantum system (a particle in a potential well, for example, an electron in the Coulomb potential of a proton) can have only strictly established energies and can occupy corresponding states described by the eigenfunctions. The modulus squared of these functions describes the spatial distribution of probability of finding the particle. It should be added that in quantum mechanics all physical quantities are represented by operators having certain mathematical properties, and the associated eigenvalue problems lead to eigenvalues (possible results of the measurement) and corresponding states. The measurement leads to a collapse of a quantum state into an eigenstate of a given quantity.

Two facts should be pointed out. First, the time-dependent part of the wave function, although it has been separated out, is still present in the full solution, but it does not affect the probability distribution of and eigenstate since its modulus squared equals 1. However, the situation changes if we consider a state being a superposition of a few eigenstates. Then we must not forget about time-dependent parts, and their presence results in time evolution of the probability distribution. The second issue is the normalisation of the wave function, which is necessary since its modulus squared multiplied by the volume element is the probability, and the probability of finding a particle overall must be equal to 1. From mathematical point of view this means that the integral of the modulus squared over the whole considered space must be equal to 1. Such normalisation is always possible since the Schrödinger equation is linear, that is any function being its solution when multiplied by a number still remains the solution.

1.2 Problem: Eigenenergies and Eigenfunctions in Rectangular Finite Quantum Well

In this project we will use the stationary Schrödinger equation (1.1.2) to find eigenvalues and eigenstates of an electron in a rectangular finite quantum well. This is not purely an academic problem since such systems are used to model, for example, semiconductor heterostructures. We describe the system as quasi one-dimensional because the changes of important physical characteristics appear in one direction only. Here, however, we will treat the quantum well as the simplest possible 1D model of the hydrogen atom. Thus Eq. 1.1.2 takes the form

$$\left[-\frac{\hbar^2}{2m_e}\frac{d^2}{dx^2} + V(x) \right] \psi(x) = \varepsilon\psi(x), \tag{1.2.1}$$

where potential is equal (Figure 1.1)

$$V(x) = \begin{cases} -V_0 & \text{if } -a/2 \le x \le a/2, \\ 0 & \text{if } x < -a/2 \text{ or } x > a/2. \end{cases}$$

In Hartree atomic units, $\hbar = m_e = e = 1$

$$\left[\frac{d^2}{dx^2} + k^2(x) \right] \psi(x) = 0, \tag{1.2.2}$$

where $k^2(x) = 2(\varepsilon - V(x))$.

The analytical solutions fall into three categories, two inside the well (Figure 1.1), which differ in symmetry (even and odd), and the third category are the corresponding solutions outside the well

$$\psi(x) = \begin{cases} A\cos(kx) & \text{for } -a/2 \le x \le a/2 \quad (\textit{even}), \\ A\sin(kx) & \text{for } -a/2 \le x \le a/2 \quad (\textit{odd}), \\ B\exp(\mp\kappa x) & \text{for } x < -a/2 \text{ or } x > a/2. \end{cases} \tag{1.2.3}$$

As one can see, the solutions are parametrised by two parameters: k – the wave number (inside the well) and κ – the rate of exponential decrease (outside). It will be shown in the next section

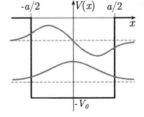

Figure 1.1 A potential well and its solution: the even (lower) and the odd (upper)

that the numerical method will consist of finding the values of these parameters for consecutive eigenstates (thus they will become indexed).

It is worth noting that the eigenfunction (thus also its modulus squared) has finite values outside the well, that is in the region where the kinetic energy of electron is negative. In classical physics a particle must not have negative kinetic energy and that is why we call such a region 'classically forbidden' and the phenomenon 'quantum tunnelling'.

1.3 Numerical Methods: Finding Roots of Characteristic Functions

The condition for the eigenvalue is that the two solutions (inside and outside the region of the well) must join smoothly (Figure 1.2), that is they must have equal values and equal values of their first derivatives at $a/2$ (because of the symmetry of the system it is sufficient to consider only one border). Thus, we have, for even solutions

$$\begin{cases} \pm A\cos(ka/2) = \pm B\exp(-\kappa a/2), \\ \mp Ak\sin(ka/2) = \mp B\kappa\exp(-\kappa a/2), \end{cases} \qquad (1.3.1)$$

and for odd solutions

$$\begin{cases} \pm A\sin(ka/2) = \pm B\exp(-\kappa a/2), \\ \pm Ak\cos(ka/2) = \mp B\kappa\exp(-\kappa a/2), \end{cases} \qquad (1.3.2)$$

where $k = \sqrt{2(\varepsilon + V_o)}$ and $\kappa = \sqrt{-2\varepsilon}$.

Dividing the first equation by the second one in the above systems, we obtain two conditions, for even (symmetric) and odd (antisymmetric) solutions:

$$\begin{cases} F_{even}(\varepsilon) = \sin(ka/2) - \kappa/k \cdot \cos(ka/2) = 0 \quad (even), \\ F_{odd}(\varepsilon) = \sin(ka/2) + k/\kappa \cdot \cos(ka/2) = 0 \quad (odd). \end{cases} \qquad (1.3.3)$$

The eigenvalues ε are found by solving these equations.

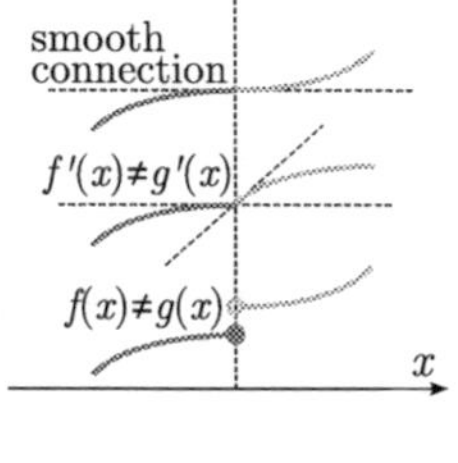

smooth connection

$f'(x) \neq g'(x)$

$f(x) \neq g(x)$

x

Figure 1.2 The solutions inside and outside the well ($f(x)$ and $g(x)$, respectively) must have same values and equal derivatives at the well border

1.4 Exercises

Obligatory

1. Using the QWELL code, tabulate functions $F_{even}(\varepsilon)$ and $F_{odd}(\varepsilon)$ These functions correspond to even and odd solutions, respectively, whose zeros are energies of quantum levels. Visualise the functions $F_{even}(\varepsilon)$ and $F_{odd}(\varepsilon)$ in one figure. Repeat the calculation and visualisation of $F_{even}(\varepsilon)$ and $F_{odd}(\varepsilon)$ for three significantly different values of the well parameters, a and V_0 (e.g. wide and shallow well, deep and narrow, intermediate).

2. (Square finite quantum well as a model of the hydrogen atom). Try to fit the first two energy levels to the ones of the hydrogen atom through variation of the parameters a and V_0, by a trial and error method. (Hint: In the beginning set the values $a = 3Bohr$ and $V_0 = 1Hartree$.) What is the value of the third energy level? One might try also to fit the first and the third levels. What would be the value of the second level then? Would it be very different from the true value? (Note that in atomic units the energy levels should be $\varepsilon_n = -1/(2n^2)$; since the well is a two-parameter system, it should be possible, in principle, to fit any two levels.)

Challenge

1. Try to construct an algorithm and write a code which automatically finds the parameters of a quantum well with energy levels close to those of the hydrogen atom with arbitrarily low uncertainty.

2. For the found eigenenergies plot the corresponding eigenfunctions and their moduli squared, with the picture of the well in background (do not normalise the functions). Note the effect of quantum tunnelling.

Project 2
Diffraction of Light on a Slit

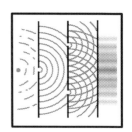

By working on this project, students will learn numerical differentiation and quadrature procedures. In particular, this project discusses the important issue of convergence with respect to the grid parameter. The numerical quadrature procedure is used to construct the DIFFRAC-TION code, which simulates diffraction of a scalar wave by a single infinite slit and a system of parallel infinite slits. The code then serves for studying the physical properties of the system.

2.1 Physics Background: Elements of Wave Physics

One can look at a wave as time- and space-dependent variation of a certain physical quantity (pressure, stress in material, displacement of an atom from its equilibrium position, etc.). It is described by a function $\varphi(t, r)$, on which, from a mathematical point of view, special conditions for differentiability are imposed. Some waves require a medium (mechanical), while others do not (electromagnetic, gravitational). The amplitudes in some waves are scalars (like in an acoustic wave), while in others they are vectors (like in an electromagnetic wave). The differential equation for the so-called linear regime (when the response is a linear function of perturbation, like change in volume vs. pressure in air, strain vs. stress in material, or the force acting on an atom in crystal vs. its displacement from equilibrium) is shown in Project 6. Here, we will focus on a certain aspect of wave physics only – the superposition principle which is a consequence of a linear character of the wave equations. Namely, if two waves φ_1 and φ_2 are solutions

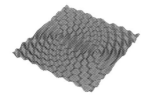

Figure 2.1 We know the principle of superposition from the water waves propagation

Figure 2.2 The intensity versus the phase difference of two superimposed harmonic oscillations

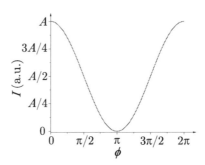

to wave equation, then also their superposition $A\varphi = \varphi_1 + B\varphi_2$ is (A, B – any variables). Using this principle it is possible to analyse the phenomena connected with the superimposing waves (Figure 2.1), like interference.

We begin with considering an abstract situation when two scalar, harmonic (sinusoidal) oscillations of the same frequency ω and amplitude A, but differing in phase by ϕ, are superimposed in some point in space. We will use the Euler representation of a complex number (see Appendix A.1) in which, for example, the real part describes physical reality. The result of superposition is

$$\varphi(t) = Ae^{(i\omega t)} + Ae^{i(\omega t + \phi)}. \tag{2.1.1}$$

Observations of wave phenomena (e.g. what we see or what we hear) are connected with the wave energy, which is proportional to the amplitude squared. Let us call it intensity (of light or sound) I. Using Equation (2.1.1), after simple algebra we find (Figure 2.2)

$$I = |\varphi(t)|^2 = \varphi(t)\varphi(t)^* = 2A^2[1 + \cos(\phi)], \tag{2.1.2}$$

where the symbol '*' denotes the complex conjugate.

When the effect of the superposition is stable, we call it interference. However, an important condition has to be fulfilled for it to be stable, namely the phase difference ϕ must not vary with time. This condition, combined with the requirement of same frequency form more general condition which we call coherence. In real systems, achieving coherence may be a big problem; for example, in the famous Young's experiment with light diffraction on a double slit, the complexity of the experimental set-up (filter, single slit, double slit) was forced just by the coherence requirement. At present, achieving coherence is one of the main challenges and scientific problems in applications of quantum systems (quantum cryptography, quantum computer).

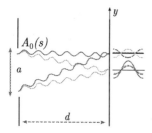

Figure 2.3 Constructive versus destructive interference in a single slit diffraction, a schematic view

The obtained formula for intensity is fundamental and general, and it shows that the effect of interference of two oscillations depends on the phase difference ϕ. There are two types of characteristic extrema: maxima for $\phi = 0 \pm n2\pi$ (n is an integer number) corresponding to constructive interference, and minima for $\phi = \pi \pm n2\pi$ corresponding to destructive interference (zeroth intensity). Besides, there is a big range of intermediate intensities.

Passing to real systems, that is the situation where oscillations depend not only on time but also on spatial variable, we should ask the question what can be the reason for the phase difference ϕ at a certain point in space. Most often the reason is the difference in so-called optical paths, that is the distance from a reference point, of known phase, for each beam, expressed in wave lengths λ. For example, for two point sources of spherical waves the phase difference is connected with the difference in distances from the sources $\phi = k(r_2 - r_1)$, where k is the wave number $k = 2\pi/\lambda$. This is exactly what happens in the Young's experiment (Figure 2.3), which in the history of science became the first irrefutable confirmation of the wave nature of light (although at this time nobody knew what kind of wave is this). A certain funny paradox is that the proof of the wave nature of light was always observed by people, in the form of coloured patterns appearing on spots of oil spilled on the surface of water. The phenomenon is a result of wave interference by division of amplitude (unlike in the Young's experiment, which is based on the division of wavefront), where the phase difference is caused by the difference in optical paths of the part reflected from the upper oil surface and the other part reflected from the oil–water interface. The coherence is most often assured here since the thickness of the oil film is usually smaller than the so-called coherence length of the light beam, that is the length within which the phase differences between chosen points do not depend on time. Presumably, the colours on the oil spots were a great puzzle for people until their origin has been explained. It is worth mentioning that the interference by division of amplitude finds presently significant application in interferometers, that is instruments for precise measurement of length. The Michelson interferometer is an example; its role in the history of science cannot be overvalued, and it was used to disprove the theory of Ether and to detect gravitational waves.

2.2 Problem: Diffraction of a Wave on a Slit

We will apply the superposition principle discussed in the previous section to construct a virtual system of wave diffraction on a certain aperture. A particular case will be an infinite slit with parallel edges. Using this virtual system it will be possible to investigate into the diffraction phenomena on such a slit at various configurations (width of the slit, the distance of the screen). Thus, we consider a plane wave of stabilised phase on a wavefront, falling on a certain aperture. According to Huygens' principle, we can treat the region of the aperture as an infinite and continuous set of point sources of spherical waves. The diffraction phenomenon is a result of the superposition of waves emitted by these sources. To be more exact, the diffraction itself is associated more with the wave deflection, that is, the fact that there is never a sharp shadow of the aperture on the screen. This is because the spherical waves emitted by point sources propagate into the whole space and not along straight lines. We will observe this for the case of a very narrow slit, whose width is much smaller than the wavelength. When additionally the interference takes place, characteristic patterns of brighter and darker regions appear. Generally, the term diffraction is used to denote both the deflection and interference at the same time. Coming back to calculus, we have to sum up the waves emitted by all the point sources, which for the case of their continuous distribution denotes quadrature. We get the diffraction integral

$$D = \int_{\text{Source}} \frac{A_0(s)}{r} \exp\left(-ikr + \phi(s)\right) ds, \qquad (2.2.1)$$

where $A_0(s)$ is the amplitude at the elementary source (or more exactly the amplitude density), $\phi(s)$ is the initial phase (at the elementary source), $k = 2\pi/\lambda$ is the wave number, r is the distance from the elementary source to the observation point, and $(A_0(s)/r)\exp\left(-ikr + \phi(s)\right)$ is the complex amplitude.

It should be noted that the $1/r$ factor holds for a spherical wave, for a cylindrical one (the case considered here) it should be replaced by $1/\sqrt{r}$ because we deal here with cylindrical waves. It should also be explained why there is no time in Eq. 2.2.1, if it is to represent a superposition of elementary waves. This is because in Euler representation of an elementary wave, $\exp[i(\omega t - kr)]$, the time-dependent factor is identical for all waves and can be factored out of the integral. When the intensity is calculated (as a modulus squared of the diffraction integral) its contribution is equal to 1. The intensity (which forms the diffraction pattern) is given by

$$I = |D|^2 = Re(D)^2 + Im(D)^2. \qquad (2.2.2)$$

In the case of an infinite slit of width a, the problem becomes two-dimensional, in the sense that no quantity varies in the direction parallel to the slit, so only two remaining directions are of interest. Assuming that the initial amplitude and phase (A_0, ϕ) are constant along the slit, the diffraction integral on the screen placed at the distance d takes the form

$$D = \int_{-a/2}^{a/2} A(r) \exp(-ikr)dx, \qquad (2.2.3)$$

where $r = \sqrt{(y-x)^2 + d^2}$, $A(r) = A_0/\sqrt{r}$ (for a cylindrical wave emitted by an elementary line source), A_0 is the amplitude at the source, x is the coordinate of the elementary source, and y is the coordinate of the observation point on the screen.

2.3 Numerical Methods: Schemes Based on Local Approximations of a Function

2.3.1 Derivatives: 2, 3, and 5-Point Schemes

Suppose we want to numerically calculate derivatives of various orders of a function $f(x)$ at a certain point x. We start from power series expansions

$$\begin{cases} f(x \pm h) = f(x) \pm f'(x)h + \frac{1}{2}f''(x)h^2 \pm \frac{1}{6}f'''(x)h^3 + O(h^4), \\ f(x \pm 2h) = f(x) \pm f'(x)2h + \frac{1}{2}f''(x)(2h)^2 \pm \frac{1}{6}f'''(x)(2h)^3 + O(h^4), \\ \cdots, \end{cases}$$

$$(2.3.1)$$

where h is a small grid parameter and $O(h^4)$ stands for the remainder of the series in which the leading (the biggest) term contains h^4. The above power series expansions, when truncated at certain order, form a system of linear equations whose unknowns are subsequent derivatives.

For example, if only the terms up to second order are preserved we get

Figure 2.4 3-Point scheme

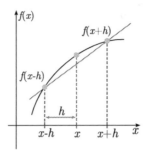

$$\begin{cases} f(x + h) = f(x) + f'(x)h + \frac{1}{2}f''(x)h^2 + O(h^3), \\ f(x - h) = f(x) - f'(x)h + \frac{1}{2}f''(x)h^2 + O(h^3), \end{cases} \qquad (2.3.2)$$

where the symbol $O(h^3)$ means that the leading term in the remainder of the series is of the order of h^3.

By subtracting or adding these equations (2.3.2), we can easily derive the 3-point schemes for the first and the second derivatives, respectively,

$$f'(x) = \frac{f(x + h) - f(x - h)}{2h} + O(h^2), \qquad (2.3.3)$$

$$f''(x) = \frac{f(x + h) + f(x - h) - 2f(x)}{h^2} + O(h^2). \qquad (2.3.4)$$

Note that the uncertainty of the second derivative is of the order of h^2; this is because the leading terms of the third order in the two expansions cancel out. The above formulas are called 3-point schemes because a local approximation of the function based on three points (parabolic) is used. The formulas give exact values of derivatives for quadratic functions. It is also interesting to note that the first derivative formula uses only two points, in spite of the fact that this is a 3-point scheme (Figure 2.4).

Using the above procedure it is easy to derive the schemes based on a larger number of points (higher order polynomial approximations) as well as derivatives of a higher order. It is also easy to incorporate a non-uniform grid (h is different at each step), although then the formulas become more complicated. It should also be noted that with high precision of representation of a number it is favourable to increase the accuracy of the derivatives simply by applying a smaller grid parameter h rather than using higher-order schemes, since higher-order schemes involve more evaluations of a function, which increases the computation time.

2.3.2 Quadrature: Rectangle, Trapezoid, and Simpson's Methods

The schemes presented here are based on the assumption that the function $f(x)$ which we want to integrate is tabulated, meaning its values are known at some points of the domain uniformly distributed along the integration interval. The grid parameter h is the distance between subsequent values of the argument in the grid $h = x_{i+1} - x_i$. The idea behind the construction of the quadrature schemes is simple. Once we can numerically calculate the derivatives of any order at the points of the grid, the function can be approximated by a polynomial of any order along certain local intervals. Thus, it can also be analytically integrated along the same local interval (Figure 2.5). The quadrature along the whole interval of interest is obtained by summing up the values of local integrals. For example by preserving only the constant in the expansion we get the 'rectangle method' (see Figure 2.5(a)), and preserving only the constant and the linear term leads to the 'trapezoid method' (see Figure 2.5(b)).

A very popular Simpson's algorithm is based on the 3-point schemes for the first and the second derivatives given in Section 2.3.1, and local quadratic approximation of the function (Figure 2.6)

$$D = \int_{-h}^{h} f(\zeta)d\zeta = \frac{h}{3}(f(x_i + h) + 4f(x_i) + f(x_i - h)) + O(h^5), \quad (2.3.5)$$

where $\zeta = x - x_i$ is a local integration variable.

Note that the local uncertainty of the quadrature is very low (of the order of h^5). It should be noted that the global uncertainty (after summation of local contributions) can increase because the local uncertainties may also accumulate (in worst case); however it will never be higher than $N \cdot O(h^5) = 1/h \cdot O(h^5) = O(h^4)$ (for the quadrature interval

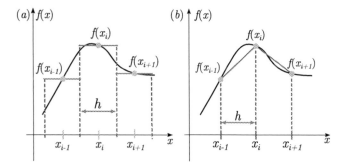

Figure 2.5 Simple quadrature methods: (a) Rectangle, (b) Trapezoid

Figure 2.6 Simpson's method

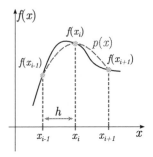

equal to 1 the lattice parameter $h = 1/N$). The strategy presented here can be used for the construction of higher-order schemes. As opposed to derivatives, in this case it can indeed lead to better accuracy at similar effectiveness, because at the same number of the $f(x)$ function evaluation its local approximation is more precise. On the other hand, decreasing the value of h (to get better local approximation) requires a larger number of evaluations. One should remember, however, that the application of higher-order polynomials always involves the risk of unwanted local behaviour of the approximating function.

2.4 Exercises

Numerical Procedures

Obligatory

1. Derive the formulas in 2.3.3 and 2.3.4 and explain why their uncertainties are $O(h^2)$.
2. (Testing the DERIV code) Substitute your own function in the segment FUNC and calculate its derivatives. Compare the results with analytical values.
3. Check the convergence of the first and the second derivatives with respect to the grid parameter h (draw derivatives as functions of $-log_{10}(h)$). Perform the tests for a single and double precision of real numbers. Discuss the results.
4. Derive the formula in Eq. 2.3.5 and explain why its uncertainty is $O(h^5)$.
5. Test the QUADRAT code by integrating a function whose analytical quadrature is known.
6. Check the convergence of the Simpson's algorithm with respect to the grid parameter h. How does it compare with the derivative

convergence? Pay attention on the computation time. Discuss your observations.

7. Estimate the number π from the length of an arc or from the surface area of a semicircle using the QUADRAT code. What precision can be achieved? Compare it with the previous exercises concerning the evaluation of π.

Supplementary

1. Derive the 5-point finite difference formulas for the first and the second derivatives of a function. Construct the functions *fp5* and *fpp5* (the first and the second derivatives with the use of the 5-point scheme) in the DERIV code and repeat Ex. 2. Compare the convergence of 3-point and the 5-point schemes.
2. Substitute second- and fifth-order polynomials in the segment FUNC and test the convergence with respect to *h*. Discuss the results.

Diffraction

Obligatory

1. Test the DIFFRACTION code by comparing the first minimum position in the far field with the analytical value. (Hint: The far-field condition can be recognised by the fact that intensities at all the minima are zero; the position of the first minimum should obey the condition $\lambda/a \approx y/d$.) One should assume the wavelength $\lambda = 1$ because the results scale with the wavelength (are qualitatively the same e.g. for optical waves and for microwaves, if only the proportions between system dimensions and the wavelength are the same).
2. Evaluate the diffraction patterns for different physical conditions: near ($d \approx a$) and far ($d \gg a$) fields, very narrow ($a \ll \lambda$), narrow ($a \approx \lambda$), and wide ($a \gg \lambda$) slits. Try to interpret the results. Discuss the effect of computational parameters on the results, and how to check whether the obtained result is physical or affected by numerical effects.
3. Modify the code so that it could calculate the diffraction pattern produced by a system of two slits. In the far field one should expect the pattern presented in Figure 2.7. Does the position of the first-order maximum agree with the theoretical prediction? The single-slit pattern, after some renormalisation, forms an envelope for

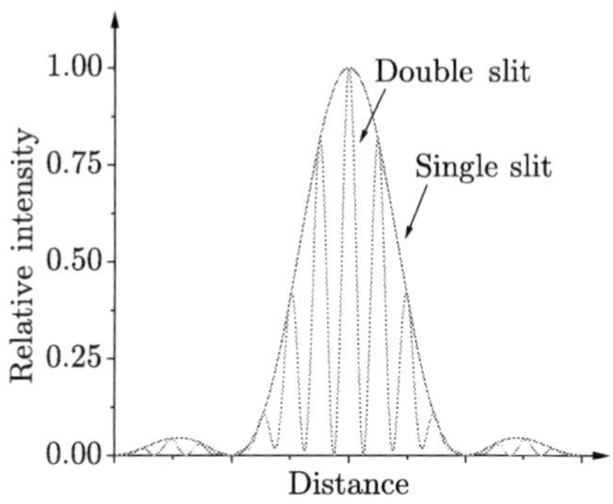

Figure 2.7 An example of the diffraction patterns in the far field for a single and a double slit. In both cases the parameters of a single slit are the same. The patterns have been normalised to 1.0 at maximum (the true value of the double slit intensity is four times higher than the single slit one)

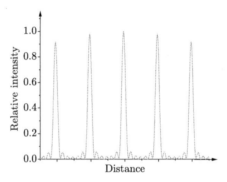

Figure 2.8 An example of pattern for diffraction grating

the double-slit pattern. Explain why and what is the renormalisation factor.

Supplementary

1. Modify the program to calculate the diffraction pattern produced by a system of many parallel slits (Figure 2.8); use a loop for that purpose. Use the modified program to simulate a diffraction grating and draw graphs of diffraction patterns in the far field for different grating parameters. How does the position of the first maximum depend on the grating parameters?

Challenge

1. Consider various initial phase distributions $\Phi(s)$ across a single slit (or at slits in diffraction grating). How do they affect the diffraction pattern? (In this task there is an analogy to GSM transmitting antenna, being usually a system of radiators, and their radiation patterns are shaped by exiting the radiators with different initial phases.)

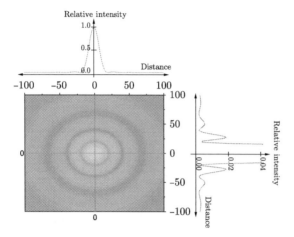

Figure 2.9 An example of diffraction pattern due to circular aperture (Airy's disc). Colour scale has been modified to enhance the outer rings

2. Consider an irregular diffraction grating in the form of periodically repeated slits of two different widths. How many parameters decide the properties of such system. Check the influence of the parameters on the diffraction pattern in the far field.

3. Consider diffraction due to a circular aperture or an aperture of an arbitrary shape. Try to construct respective program. (Hint: Instead of integration a summation over a discrete points of elementary sources should be applied.) See Figure 2.9 for reference.

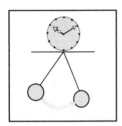

Project 3
Pendulum as a Standard
of the Unit of Time

This project is devoted to the Initial Value Problem (IVP) for ordinary differential equations. Here, students will learn various multipoint recursion schemes, apply them to examples of equations whose analytical solutions are known, check the convergence with respect to the grid parameter, and compare the effectiveness of the schemes. Finally, a chosen scheme is applied to study the properties of the compound pendulum, in particular the dependence of the period of oscillation on energy. The results may serve to discuss the applicability of the pendulum as a standard of the unit of time.

3.1 Physics Background: Newton's Laws of Motion, Equation of Motion

This project is a good opportunity to recall and discuss briefly Newton's laws of motion, published in 1687 in the famous work *Philosophiae Naturalis Principia Mathematica*. This work, in spite of concepts, language, and mathematical apparatus being far from modern standards, became a foundation of physics and the whole science. The first law of motion, although at first sight looks very simple and seems to be a consequence of the second law of motion, is in fact the first fundamental physics law discovered by people. It says that an isolated body (body not subjected to any external interactions) does not change the state of its motion, that is will either stay at rest or keep moving at constant velocity. This law should be treated as a postulate, since it cannot be experimentally verified on the Earth (nor in fact anywhere else). According to our everyday experience, bodies in motion will sometimes stop, and such a view was obligatory over ages (Aristotle) before Newton presented his first law of motion. An important aspect of Newton's first law is that it can be used to introduce the concept of inertial reference frame, as the one in which we could hypothetically observe the described phenomenon. It is also worth mentioning that in the formulation of the law we do not have to use the concept of force (not defined yet). Unfortunately, in many textbooks force is used in its formulation ('no force is acting' or 'all the forces cancel out'), maybe following Newton himself, who in his work also used the concept of force to express the first law. However, the richness of concepts in modern science (e.g. an isolated system, postulate) allows for the formulation which is formally correct and internally uniform. In Newton's second law the concept of force appears as the reason for the change in the state of motion. Force can be defined via acceleration, which is a well-established kinematic quantity. At this point, the fact should be taken into account that the change in the state of motion is determined by certain internal characteristics of the body – inertia, measured by a quantity called mass. Until recently, in the SI system of units, the unit of mass, 1 kilogram, was defined by a physical standard. At present, the SI system uses universal physical constants to define units. Next, if we know already the unit of mass, then the measure of force is an acceleration of a body of unit mass under the action of this force. Now, we can formulate Newton's second law: if on a body acts a force $\bar{F}$, being a vector sum of all the forces present, then its instantaneous acceleration $\bar{a}$ is directly proportional to the force and inversely proportional to its mass. Finally, Newton's third law says that in an inertial reference frame the forces appear as a consequence of interactions between bodies, and the interactions are always mutual, that is if a body A acts on a body B with a certain force, then body B also acts on body A with the force of the same value but opposite direction.

The equation of motion is just Newton's second law expressed in the form of differential equation

$$\frac{d^2 r}{dt^2} = \frac{F(t, r, v)}{m},\qquad(3.1.1)$$

where $\bar{r}$ is the position vector and $\bar{v}$ velocity vector.

This is the second-order non-uniform differential equation, being most often the basis of the so-called IVP. In such problems we want to find the evolution of trajectory in phase space (position and momentum) of systems at given initial conditions. This will also be a basis of the present and the following projects.

3.2 Problem: Simple Pendulum as a Standard of the Unit of Time

We will apply Newton's second law for a simple pendulum – a point mas suspended from a massless, stiff rod of length l. The uniform (independent of position) vertical force, whose origin is gravitation (will be

Figure 3.1 The
mathematical model of
pendulum

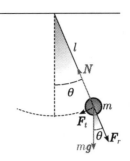

discussed in the next project), acts on the point mass. The force and
its convenient decomposition is shown in Figure 3.1. The gravitational
force has a component along the rod F_r, which cancels out with the
force due tension in the rod N. Another component is tangent to the
circle drawn by the moving point mass and is a cause of motion. The
equation of motion Eq. 3.1.1 takes the form

$$\frac{d^2s}{dt^2} = -g\sin(\theta), \qquad (3.2.1)$$

where s is a coordinate of the mass position along the circle, measured
from the lowest point.

When we divide both sides of the equation by the length l, and note
that the ratio of the arc length and the radius of the circle is just an angle
in arc measure, we get

$$\frac{d^2\theta}{dt^2} = -\frac{g}{l}\sin(\theta), \qquad (3.2.2)$$

where $\theta = s/l$ is the swing angle and g the free fall acceleration.

For small swing angles $\sin(\theta) \approx \theta$, the oscillation is harmonic

$$\frac{d^2\theta}{dt^2} = -\frac{g}{l}\theta, \qquad (3.2.3)$$

$$\theta(t) = \theta_0 \sin(\Omega t), \qquad (3.2.4)$$

where θ_0 is the swing angle amplitude, and $\Omega = 2\pi/T_0 = \sqrt{g/l}$.

The period of harmonic oscillations T_0 does not depend on the
amplitude. This phenomenon (isochronism) allows us to use a pendulum as the standard of the unit of time in pendulum clocks (the first
one constructed in 1656 by Christiaan Huygens). However, for bigger
angular amplitudes the period begins to change and to find it, it is necessary to numerically solve the differential equation (3.2.2) for given
initial values of angle and their first derivatives (IVP).

A convenient way of setting the initial state of the pendulum is to use its total energy $E = mgl(1 - \cos(\theta)) + ml^2\omega^2/2$ expressed in units of the maximum potential energy (with respect to the lowest position), that is $\epsilon = E/2mgl$. Then we expect the period to be constant for $\epsilon \ll 1$ (isochronism), tend to infinity for ϵ approaching 1, and to behave like $1/\sqrt{\epsilon}$ for $\epsilon \gg 1$ where the rotational energy prevails.

3.3 Numerical Methods: Recursive Methods Based on Local Extrapolation of One-Step Integral Integrand

A system of first order, linear, non-uniform differential equation reads

$$\left\{ \frac{dy_i}{dx} = f(y_1, \ldots, y_N, x), \right. \tag{3.3.1}$$

where x is the independent variable, and $\{y_1, \ldots, y_N\}$ are the dependent variables (functions of x).

Linear differential equations of higher order can be decomposed into a system of equations (3.3.1) by defining auxiliary functions as derivatives of the function $y(x)$.

Developing the numerical methods for a single equation

$$\frac{dy}{dx} = f(y, x) \tag{3.3.2}$$

is sufficient, since the methods can be easily adopted to the system (3.3.1).

The algorithms presented here are based on discretisation of the independent variables. A regular mesh (grid) of points $\{x_0, x_1, \ldots, x_N\}$, in the interval (x_0, x_N), is introduced. The distance between neighbouring points $h = (x_{i+1} - x_i)$, also $h = (x_N - x_0)/N$, is called the grid parameter h. We seek a recursion formula of the form

$$y_{n+1} = F(y_n, y_{n-1}, y_{n-2}, \ldots), \tag{3.3.3}$$

which would allow the function $y(x)$ to evolve, beginning from the initial point $y_0(x_0)$.

A starting formula is an exact integral of the equation (3.3.2) over the interval (x_n, x_{n+1})

$$y_{n+1} = y_n + \int_{x_n}^{x_{n+1}} f(x, y)dx; \tag{3.3.4}$$

Figure 3.2 Euler's method

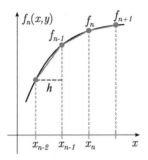

let us call the integral appearing in Eq. (3.3.4) (exact) formula a one-step integral.

In general, the explicit form of $f(x, y)$ is not known since the $y(x)$ is not known (it is to be found). However, values of $y(x)$ at previous points of mesh $y_n, y_{n-1}, y_{n-2}, \ldots$ are known and they can be used to extrapolate $f(x)$ over the interval (x_n, x_{n+1}), which will allow us to perform the one-step integration. Depending on the number of points used in the extrapolation we get the so-called explicit schemes of various orders. For example, the most simple Euler's formula (Figure 3.2) (based on the extrapolation by a step function) reads

$$y_{n+1} = y_n + f_n h + O(h^2), \tag{3.3.5}$$

where $f_n = f(x_n, y_n)$. The leading term of the local deviation from the exact solution is on the order of h^2. It should be noted that after N steps the uncertainty (global) will be one order lower because $h \simeq 1/N$, thus $h^2 \cdot N = h^2/h = h$.

The three step Adams–Bashforth scheme, based on the parabolic extrapolation, has the form

$$y_{n+1} = y_n + \frac{h}{12}(5f_{n-2} - 16f_{n-1} + 23f_n) + O(h^4). \tag{3.3.6}$$

3.3.1 Runge–Kutta Methods

The Runge–Kutta methods are based on more advanced mathematical concepts which will not be discussed here, although the integral (3.3.4) is still used as the starting point. They are regarded as the best integration schemes, whose great advantage is the fact that only one, already known, point x_n, y_n (e.g. the initial condition) is needed to perform the integration step, whereas the multipoint recursion schemes cannot be started from a single initial condition. As an example, the fourth-order

Runge–Kutta method has the form

$$
\begin{aligned}
k_1 &= hf(x_n, y_n), \\
k_2 &= hf(x_n + 1/2h, y_n + 1/2k_1), \\
k_3 &= hf(x_n + 1/2h, y_n + 1/2k_2), \\
k_4 &= hf(x_n + h, y_n + k_3), \\
y_{n+1} &= y_n + \tfrac{1}{6}(k_1 + 2k_2 + 2k_3 + k_4) + O(h^6).
\end{aligned}
\tag{3.3.7}
$$

The unknown function $y(x)$ appearing in the integrand (3.3.4) can be also approximated by a polynomial of any order. In this case the unknown value y_{n+1} appears on the right-hand side of the formula as well and has to be evaluated by solving the equation. This strategy leads to a group of the so-called implicit schemes. They are most often used in the predictor-corrector methods in which a predicted value of y_{n+1} found from one of explicit schemes is then corrected by the implicit scheme of higher order (by substituting the predicted value on the right-hand side of the equation).

3.4 Exercises

Numerical Procedures
Obligatory

1. Derive Eq. 3.3.6.
2. Use the IVP code to evaluate the function which is a solution to the differential equation $dy/dx = y, y(0) = 1$, with the use of three methods: Euler, third-order Adams–Bashforth, and fourth-order Runge–Kutta. Visualise the results, together with the analytic solution $y = e^x$.
3. Modify the code so that it gives at the output the $y(1)$ value as a function of the grid parameter $-\log_{10}(h)$ (the logarithmic scale). Check the convergence of the three schemes with respect to the grid parameter. Classify the schemes with respect to their quality and efficiency.
4. Look at the construction of the IVP2D code and note how the 1D algorithms are adopted to solve the second-order differential equation. Use the code to evaluate the function being a solution to the harmonic oscillator equation $d^2y/dx^2 = -k \cdot y$. Draw the function $y(x)$ and interpret the result (Figure 3.3(a)). Add a line to the code that calculates the total energy of the oscillator and output the result. Find the range of the grid parameter at which the energy is conserved. Repeat the calculations for Euler and Runge–Kutta methods.

Figure 3.3 Different cases of oscillators motion: (a) Harmonic oscillator, (b) Damped harmonic oscillator, and (c) Driven harmonic oscillator

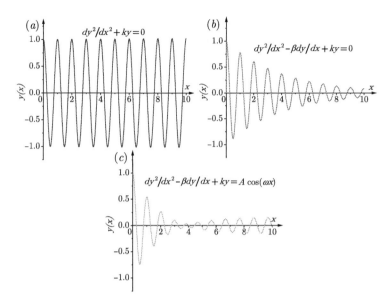

Supplementary

1. Modify the IVP2D code so that it could solve the equations:
   (a) $d^2y/dx^2 - \beta \cdot dy/dx + k \cdot y = 0$ (damped oscillator Figure 3.3(b)). For the chosen parameters elastic constant, mass, damping constant, and a proper time-step (grid parameter), perform the calculations.
   (b) $d^2y/dx^2 - \beta \cdot dy/dx + k \cdot y = A \sin(\omega \cdot x)$ (driven oscillator Figure 3.3(c)) for different driving frequencies. Draw resulting functions $y(x)$ and identify the time ranges of unstable and stable oscillations. Observe and discuss the behaviour of oscillation energy (kinetic plus potential) as a function of time. Try to find a few points on the resonance curve, close to maximum (near the resonant frequency), and much lower and higher frequencies. Note that for higher frequencies a smaller time-step may be needed (how to check it?).

Challenge

1. Consider the case of coupled harmonic oscillators described in Project 9. Read the project and follow the exercises given there.

Compound Pendulum

1. Test the PENDULUM code by comparing the period of small oscillations with the analytic solution.

2. Perform a few calculations for different total energies of the pendulum (the total energy is expressed in units of the maximum potential energy) and draw the time dependence of the angle and angular speed. Propose an algorithm for automatic calculation of the period and implement it. (Hint: For that purpose it is easier to use the angular speed rather than the angle – explain why?) Evaluate the relative deviation of the period at a given energy $T(\epsilon)$ from the analytic value (for small oscillation) $T_0 = 2\pi \sqrt{(l/g)}$, $(T(\epsilon) - T_0)/T_0$, for three energy ranges: $\epsilon = (0.001-0.01)$, $\epsilon \approx 1.0$, and $\epsilon \gg 1.0$. Draw and discuss the results.

3. Focus on a case where $\epsilon = 1.0$ and perform the calculations for different time steps (differing by a few orders of magnitude). Draw the time dependence of the angle and explain the observed differences.

Project 4
Planetary System

Molecular dynamics (MD) is a very popular type of simulation in physics. It is essentially just a procedure of solving the equations of motion for a system of many particles, either classical (Newtonian) or relativistic, from atomic to cosmic scale. A rigorous analytical solution exists only for two-particle systems (and three-particle in special cases), thus a computer simulation seems to provide the only possibility of studying such systems theoretically. Molecular dynamics simulations are very popular in studying dynamics and thermodynamics of polyatomic systems but they are also used on a cosmic scale (motion of planets, stars, galaxies). All we need to know to construct a MD code is the law of interaction between the particles and Newton's (or relativistic) equations of motion. In this project, a simple two-dimensional planetary system will be considered with two planets and a fixed star as the source of the central force. Newton's law of universal gravitation is used as the interaction law between planets and between planets and the star. The Verlet algorithm, in its simplest form (most often used in simple MD simulations), for solving the initial value problem will be applied.

4.1 Physics Background: Law of Universal Gravitation

In this project we will solve equations of motion - an initial value problem for a system of two planets moving in the field of a fixed star. As has been mentioned in introduction of this project, besides the laws of motion which have been discussed in previous project, the law of interaction between objects is needed. Here, it is the gravitational interaction presented and described by Isaac Newton in *Philosophiae naturalis principia mathematica*, quoted in the previous chapter. It says that two point masses m_1, m_2 at distance r attract each other with the force $\bar{F}$ whose value is directly proportional to the product of masses, inversely proportional to the distance squared, and the force is attractive

$$\boldsymbol{F} = -G\frac{m_1 m_2}{r^2}\hat{r}, \qquad (4.1.1)$$

where $\hat{r} = r/r$ is a unit vector along r. In this equation gravitational constant G is necessary since the units of mass and distance have been established in advance. Moreover, the multiplication of the ratio $(m_1 m_2)/r^2$ by G makes the force realistic (extremely small for 1 kilogram and 1 metre). Even if the force seems to be so small, it plays a key role on a cosmic scale and is responsible for the structure of the universe (stars, planets, galaxies, black holes, etc.). Also, owing to this force, we are safely kept on the surface of the Earth, but it creates challenges when we want to leave the Earth and travel to outer space. When considering gravitational interactions, one should always keep in mind that Newton's third law applies: it states the mutuality of interactions. A force of value given in Eq. 4.1.1 acts on both point masses along the line joining them and is attractive, that is, it always acts in the direction towards the other mass.

There are a few issues that deserve a brief discussion here. The first issue is the important superposition principle which says that the interaction between two point masses does not depend on the presence of other masses in the region. In other words, the gravitational force acting on a certain mass is a vector sum of all the gravitational forces due to other objects. This is not an obvious fact since, for example, in the system of many atoms they attract or repel each other in a complicated way, and often many-body force fields are necessary. For example, two hydrogen atoms interact in a different way when they are in the H_2 molecule and when the oxygen atom is nearby (in the water molecule H_2O).

The second issue is the principle of equivalence, being something very mysterious over two ages. It turns out that the mass responsible for the gravitational interaction is exactly the same mass which is a measure of inertia of a body. Thus, for example, no experiment can differentiate between gravitational and inertial forces within a closed room. This concept forms a fundamental basis for Albert Einstein's construction of the general theory of relativity. Developed in the early twentieth century, this represents the modern theory of gravitation.

Finally, it should be recalled that the gravitational force is conservative, which means, for example, that the work done by an external force opposite to the gravitational force along a closed path is zero. A related property of such force field is that it can be described by the potential energy $U(r)$, a scalar function of position. The relationship between the potential energy and the force is $F = -\nabla U(r)$, where ∇ is the gradient operator.

Figure 4.1 All massive objects attract each other according to Newton's law of universal gravitation

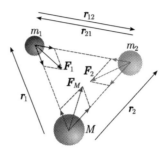

For two point masses the potential energy is given by the expression

$$U = -G\frac{m_1 m_2}{r}, \qquad (4.1.2)$$

which will be used in Section 4.3, at the reduction of the single particle motion in the field of a star to one dimension.

4.2 Problem: Motion of Planets in the Field of a Fixed Star

The forces acting on planets due to the Star fixed in the origin of a reference system and due to the other planets moving in the same plane (two-dimensional system Figure 4.1) can be expressed as

$$\mathbf{F}_1 = -G\frac{m_1 M}{r_1^2}\hat{r}_1 + G\frac{m_1 m_2}{r_{12}^2}\hat{r}_{12} \text{ and}$$

$$\mathbf{F}_2 = -G\frac{m_2 M}{r_2^2}\hat{r}_2 + G\frac{m_2 m_1}{r_{21}^2}\hat{r}_{21}, \qquad (4.2.1)$$

where $\mathbf{r}_{ij} = \mathbf{r}_j - \mathbf{r}_i$.

The Newton equations of motion for the two planets are

$$\begin{cases} d\mathbf{p}_1/dt = \mathbf{F}_1, \\ d\mathbf{r}_1/dt = \mathbf{p}_1/m_1, \\ d\mathbf{p}_2/dt = \mathbf{F}_2, \\ d\mathbf{r}_2/dt = \mathbf{p}_2/m_2. \end{cases} \qquad (4.2.2)$$

This is a system of differential equations, linear and non-uniform. They can be solved with the methods described in the previous project. However, here we will apply an algorithm which is specific for the equations of motion and often used in MD problems – the Verlet

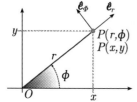

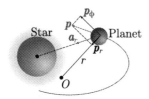

Figure 4.2 Construction of a local rectangular coordinate system for representing vectors in polar coordinates

Figure 4.3 In general the origin of the coordinate system can be placed at any point of the space (not associated with star's position)

algorithm. Its construction starts from a second-order equation, and its basic version is described in this project.

4.3 Reduction of a Single Planet Motion in a Central Field to 1D

The exercises related to the present project begin with the analysis of a single planet motion in a central field of a star. Such a motion can be reduced to one dimension, which helps to understand its properties. The transformation is presented next.

We begin with writing the classical Hamiltonian of the system (its total energy)

$$H = \frac{\mathbf{p}^2}{2m} - G\frac{mM}{r}. \tag{4.3.1}$$

If $\mathbf{p}$ is represented in polar coordinates (see Figure 4.2), $\mathbf{p} = p_\phi \hat{e}_\phi + p_r \hat{e}_r$, where $\hat{e}_\phi, \hat{e}_r$ are unit vectors of the local (attached to the particle–planet) rectangular coordinate system (Figure 4.3), then the Hamiltonian takes the form

$$H = \frac{p_r^2}{2m} + \frac{(p_\phi r)^2}{2mr^2} - G\frac{mM}{r}, \tag{4.3.2}$$

where, in the second term, the factor r^2 has been introduced in both the numerator and denominator. Since the quantity $p_\phi r$ is just the angular momentum L which is conserved in the central field, we can rewrite the Hamiltonian as follows:

$$H = \frac{p_r^2}{2m} + V_{eff}(r), \tag{4.3.3}$$

Figure 4.4 The scheme of a planet motion along an ellyptical orbit with respect to the star located in one of the foci of the ellipse, according to Kepler's first law

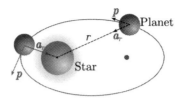

Figure 4.5 Effective potential, after reduction of 2D system to one dimension. It is the sum of two terms: the centrifugal potential $\sim 1/r^2$ and the gravitational potential $\sim -1/r$. The graph identifies planet trajectories depending on its energy

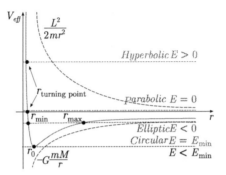

where $V_{\mathit{eff}}(r) = \frac{L^2}{2mr^2} - G\frac{mM}{r}$ is r-dependent effective potential energy of the particle in non-inertial reference frame attached to the position vector, being a sum of the centrifugal potential energy $\frac{L^2}{2mr^2}$ (present in the non-inertial reference frame) and the gravitational potential energy $-G\frac{mM}{r}$ (Figure 4.4). Once the Hamiltonian is known, the Hamilton canonical equations for 1D motion can be written, but here the effective potential (potential energy of a unit mass) V_{eff} will be of interest. For a given angular momentum (conserved) the effective potential does not change, but the total energy (also conserved) of the moving particle can have different values. Figure 4.5 shows how the character of the trajectory depends on its total energy relative to the effective potential. One can easily distinguish circular, elliptic, parabolic, and hyperbolic trajectories, which is a subject of one of the exercises in this project.

It should be mentioned here that a very similar operation (reduction of a system to 1D) is performed when the hydrogen atom is considered. In this case the quantum 3D Hamiltonian is reduced to 1D by introducing the centrifugal potential. The angular momentum is quantised, and its quantisation is represented by the so-called orbital quantum number l. The quantum number l then enters the effective potential and thus the solutions (quantum states) of the hydrogen atom.

4.4 Numerical Method: Verlet Algorithm

The Verlet algorithm is a method developed for numerically solving Newton's equations of motion for a system of particles (Molecular Dynamics), and is most often used in such simulations. Here, we will learn its simplest form and the idea of the algorithm will be presented as an example of 1D motion; however, as was the case in previously discussed algorithms, it can be easily generalised to many dimensional systems of many particles.

Newton's equation of motion of a point mass m in 1D reads

$$\frac{d^2x}{dt^2} = F/m, \tag{4.4.1}$$

where F is the force acting on a particle of mass m, and x is the coordinate of the particle on the X axis.

To study the trajectory in phase space, the velocity $v = dx/dt$ is also needed.

We use three-point numerical formulas for function derivatives (2.3.3,2.3.4) to express acceleration and velocity, which, when rearranged, lead to the following recursive schemes (Verlet algorithm):

$$x_{n+1} = 2x_n - x_{n-1} + \tau^2 F/m + O(\tau^4),$$
$$v_n = (x_{n+1} - x_{n-1})/(2\tau) + O(\tau^2), \tag{4.4.2}$$

where τ is the time step.

It should be noted that as initial conditions usually position and velocity (or momentum) are given (x_o, v_o), thus, there is certain difficulty in starting the recursive scheme since it needs two initial points. The difficulty can be overcome, for example, by assuming that over the first time step the system performs in a uniformly accelerated motion

$$x_1 = x_o + v_o\tau + (F/m)\tau^2/2 + O(\tau^3), \tag{4.4.3}$$

which, however, is less accurate by one order of magnitude of τ and thus should be applied with caution. Alternatively, more accurate schemes (e.g. Runge–Kutta) for the first time step can be used.

4.5 Exercises

Obligatory

1. (Testing the program) Test the PLANETS code on the example of a motion of a single planet along a circular orbit. For that purpose modify the code so that one of the planets is fixed far aside and its

mass is very low. Check the energy and the linear momentum conservation and find the maximum time-step for which those quantities
are conserved. Compare the results with an analytical solution.

2. (Motion of a single planet) For a given angular momentum L,
   draw the effective potential and try to identify the circular, elliptic,
   parabolic, and hyperbolic trajectories for a single planet. Perform
   simulations for each of these cases and draw the trajectories. Hint:
   To set the appropriate initial conditions for different energies you
   should use the Eq. 4.3.2 with the radial momentum set to zero $p_r = 0$

$$H = \frac{L^2}{2mr^2} - G\frac{mM}{r}. \tag{4.5.1}$$

Then, at given L the total energy becomes a function of r. However,
one should remember that to keep the angular momentum constant
when r is varied, it is necessary to change also the transversal component of the momentum, according to $p_\phi = L/r$. In practice, set r
as the x component (at $y = 0$), and $p_\phi = p_y$ (at $p_x = 0$).

3. For an elongated elliptical orbit, observe the evolution of the total
   energy, and explain the observed behaviour.

4. Include the motion of the second planet. Choose one of the suggested scenarios and perform a simulation.

   (a) Planetary system – motion of planets of different masses along
       independent orbits, in the central field of a star
   (b) Collision of planets – motion of planets along close orbits but
       in opposite directions, observation of collision and its conse-
       quences
   (c) Planet with a moon – treat one of the planets as a satellite of the
       other one, or double system of two identical planets on an orbit
       along a star
   (d) Planet in a double-star system – treat one of the planets as the
       second heavy star, and observe the motion of the second planet
       between the two stars
   (e) Double star – treat both planets as heavy stars (reduce the mass
       of the central star to a small value), and observe the system
       dynamics for different initial conditions
   (f) Your own scenario.

To be sure that the observed effects are physical and are not the result
of numerical artefacts, for each simulation it is obligatory to observe
the behaviour of energy with time (it should always be conserved)
and of the angular momentum (if it should be conserved in a given

system). If the conservation principles are not obeyed, appropriate changes in the computational parameters should be introduced.

Supplementary

1. Perturbed motion: change the mass and position of the fixed planet, and observe its effect on the trajectory of the second planet (a perturbation). Check the energy and the momentum conservation. Should both these quantities be conserved? Try to characterise the effect of the perturbation on motion of the planet.

Challenge

1. By introducing proper numerical values of the parameters observe the effect of the presence of Venus (or Mars) on the trajectory of the Earth. (Hint: Keep the position of Venus (Mars) fixed.) Note that the period of motion of the Earth is about 365 days, what should be the time-step in this case?
2. For an elliptical orbit of a single planet, check the third Kepler law of planetary motion.
3. The choice of the simulation time-step is always a compromise between the simulation time and accuracy. From that point of view the optimal choice is the biggest step which guarantees good accuracy. However, there are some situations where a constant time step leads to the waste of computer resources. An example of such situation is the case of elongated elliptical orbit simulation (see obligatory Ex. 3). At the long part of the orbit the motion is almost along a straight line, where a long time-step is sufficient. A critical point is near the star (perihelion), where there are abrupt changes in the value and the direction of velocity, and a short time-step is necessary. Setting a constant time-step would require the use of the shorter one, which would lead to a significant and useless elongation of the computation time. Moreover, in the case of many particle systems, it is difficult to predict where the short time-step is needed.

   In view of these remarks, propose an algorithm in which the time-step changes dynamically, depending on the need. (Hint: Use the behaviour of the velocity.) Implement this algorithm and test it on the example of the elongated orbit (obligatory Ex. 3).

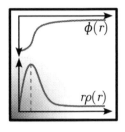

Project 5
Gravitation inside a Star

This project is devoted to the boundary value problem (BVP) for ordinary differential equations. The case of gravitational field inside a star of a model radial mass density distribution is considered. The problem is formally equivalent to the problem of the electric field inside an atom (the Hartree contribution). The partial differential equation of Poisson's type, due to high symmetry (spherical), reduces to a second-order ordinary differential equation. Unlike the initial value problem, here, two conditions necessary to uniquely identify the solution are given at two ends of the independent variable range (not at one end, like in the initial value problem). Usually, the values of the function are given at the boundaries (BVP). It is worth noting at this point that for this type of equation a whole family of functions differing by a linear function (established by two parameters of arbitrary values) form solutions. Application of boundary values eliminates all functions from this family except one. We will try to use a very accurate three-point recursive Numerov's (Cowell's) algorithm, presented later in this project. To start the three-point recursive algorithm, values of the function at two points at one end are needed. For the purpose of this project we take the second point value from the known analytic solution, but this will turn out to be insufficient since analytic solution of differential equation is not the same as the solution to a discretised equation. Thus, we face a problem: the numerical solutions wander linearly up or down and appear to be very sensitive to the value chosen for the second point. Fortunately, in the considered problem the solution function converges to a known constant in infinity which allows us to apply a recursive scheme backwards (at a sufficient distance the two initial values of the function are the same), and obtain the solution that way. However, such situations are not always the case and recursive schemes in principle cannot be applied. An interesting alternative is to treat the three-point recursive formula as a tridiagonal system of linear equations for the unknown values of the function at grid points (with the known values at both ends), and they should be solved by the Gaussian Elimination with Backward Substitution algorithm described in Project 7.

This project is also an introduction to one of the advanced projects, described in Project 11, in which the gravitation inside a star is found together with the density distribution.

5.1 Physics Background: Gauss's Law, Poisson's Equation

Gauss's law is expressed by the first four Maxwell's equations describing completely the classical electrodynamic phenomena. It can be formulated in integral or differential form. The forms are mathematically equivalent and the latter one, if additionally the relationship between the electric field and the potential is taken into account $(E = -\nabla\phi(r))$, is called the Poisson's equation. Mathematically, the integral form of Gauss law can be expressed as

$$\oint_S E(s)\,ds = AQ_w, \tag{5.1.1}$$

in other words, the total flux of electric field through a closed surface is equal to the resultant charge (i.e. considering also the signs) inside the surface. In the integrand there is a dot product of electric field and oriented surface element, which means that the projection of the electric field on the direction perpendicular to the surface is multiplied by the area of the surface element. This is an elementary contribution to the flux and the sum (integral) of these contributions will result in the total flux. According to convention, in this formulation the direction of the oriented surface element vector is outwards. Gauss's law is a direct consequence of $1/r^2$ type of interaction, which appears, for example, in electrostatics (Coulomb's law) or in universal gravitation law, and can be derived from this kind of interaction. Therefore, it can be used in both electrostatics and gravitation (as in this project), only the constants A will differ, for example in SI unit system in electrostatics $A = 1/\varepsilon_0$, where ε_0 is the permittivity, and in gravitation $A = 4\pi G$, where G is the gravitational constant. In gravitation, the role of the electric field is taken over by the gravitational acceleration a_G.

The differential form can be formally derived from the integral one, and after taking into account the Eq. 5.1.1 relationship between the electric (gravitational) potential and the electric field (gravitational potential) we get

$$\nabla^2\phi(\mathbf{r}) = -A\rho(\mathbf{r}), \tag{5.1.2}$$

where ∇^2 is, known from previous projects, Laplace's operator and $\rho(\bar{r})$ is the density of charge (mass).

Figure 5.1 Spherical coordinates

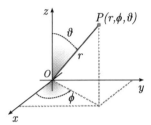

5.2 Problem: Gravitational Field Due to a Continuous Mass Density Distribution

Poisson's equation is, in general, a partial differential equation which together with the boundary conditions forms the BVP, the solution to which is the electric potential function $\phi(\mathbf{r})$ for a given charge density distribution $\rho(\mathbf{r})$.

We begin with writing the Laplacian operator in Eq. 5.1.2 in spherical coordinates (see Figure 5.1)

$$\nabla^2\phi = \frac{1}{r^2}\frac{\partial}{\partial r}\left(r^2\frac{\partial\phi}{\partial r}\right) + \frac{1}{r^2\sin\vartheta}\frac{\partial}{\partial\vartheta}\left(\sin\vartheta\frac{\partial\phi}{\partial\vartheta}\right) + \frac{1}{r^2\sin^2\vartheta}\frac{\partial\phi}{\partial\varphi}. \tag{5.2.1}$$

Since in spherical symmetry all partial derivatives of the potential with respect to angles disappear, Poisson's equation acquires a simpler 1D form

$$\frac{1}{r^2}\frac{d}{dr}\left(r^2\frac{d\phi(r)}{dr}\right) = 4\pi\rho(r). \tag{5.2.2}$$

In Eq. 5.2.2, we assumed the value of the gravitational constant $G = 1$ and omitted the minus sign, which holds for the gravitational field. A standard substitution $\phi(r) = \varphi(r)/r$ simplifies the formula even further

$$\frac{d^2\varphi(r)}{dr^2} = 4\pi r\rho(r). \tag{5.2.3}$$

This is a second-order, linear, inhomogeneous, ordinary differential equation. Such an equation needs two conditions to uniquely define the solution. In the case of an initial value problem, these are usually the value of the function and its first derivative at certain initial point, which allows to start the three-point recursive numerical algorithm (like the Verlet algorithm). However, in the case of Eq. 5.2.3, there are boundary conditions, that is values of the function at two ends of certain argument interval. This makes the application of the recursive schemes very

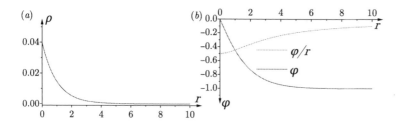

Figure 5.2 Analytical solution of Eq. 5.2.2 (a) given density of mass, and (b) gravitational potential

inconvenient because an additional condition must be found in order to use them.

We will use Eqs. 5.2.1–5.2.3 to find the gravitational potential inside a star. Suppose the mass density distribution inside a star is given by the formula (Figure 5.2(a))

$$\rho(r) = \frac{1}{8\pi}e^{-r}, \qquad (5.2.4)$$

in which case the total mass of the star is equal to 1 (the mass of the Star becomes the unit of mass)

$$M = \int \rho(\mathbf{r})d^3\mathbf{r} = \int_0^\infty \rho(r)4\pi r^2 dr = 1. \qquad (5.2.5)$$

The exact solution to this problem is (Figure 5.2(b))

$$\varphi(r) = 1 - \frac{r+2}{2}e^{-r}, \qquad (5.2.6)$$

from which $\phi(r) = \varphi/r$ follows immediately (note that specific units for the gravitational potential are used here). Through independent specu-lations it is possible to find the additional initial condition to start the recursion Numerov algorithm (described in the next section), but for the purpose of this project we will use the above formula to find the necessary conditions (e.g. values $\varphi(r = 0)$ and $\varphi(r = h)$).

5.3 Numerical Method: Numerov–Cowells Algorithm

We consider a class of the second order, linear, inhomogeneous dif-ferential equations, which describe various physical systems (including almost all discussed so far)

$$\frac{d^2y}{dx^2} + k^2(x)y = S(x), \qquad (5.3.1)$$

where $k^2(x)$ is a real function and $S(x)$ is a 'driving' term, making the equation inhomogeneous.

Using the expansions (2.3.1) one can write down

$$\frac{y_{n+1} - 2y_n + y_{n-1}}{h^2} = y_n'' + \frac{h^2}{12}y_n'''' + O(h^4). \qquad (5.3.2)$$

On the right-hand side of the above equation a second derivative of the differential equation can be applied

$$y_n'' = (-k^2 y)_n + S_n, \qquad (5.3.3)$$

and the forth derivative can be evaluated numerically using the three-point scheme as the second derivative of $y''(x)$

$$y_n'''' = -\frac{(k^2 y)_{n+1} - 2(k^2 y)_n + (k^2 y)_{n-1}}{h^2} + \frac{S_{n+1} - 2S_n + S_{n_1}}{h^2} + O(h^2),$$
$$(5.3.4)$$

which, when substituted into (5.3.2), leads to the numerical expression (Numerov–Cowell's algorithm)

$$\left(1 + \frac{h^2}{12}k_{n+1}^2\right)y_{n+1} - 2\left(1 - \frac{5h^2}{12}k_n^2\right)y_n + \left(1 + \frac{h^2}{12}k_{n-1}^2\right)y_{n-1} =$$
$$= \frac{h^2}{12}(S_{n+1} + 10S_n + S_{n-1}) + O(h^6).$$
$$(5.3.5)$$

Note that the substitution does not spoil the overall uncertainty (accuracy) $(O(h^4))$ and the final local uncertainty of the scheme remains very high $(O(h^6))$.

An important advantage of the expression is that it can be rearranged to give recursive formulas either in 'forward' or 'backward' direction, which will be used in this project. It can also be treated as a three-diagonal system of equations, which can be very effectively solved numerically with the use of 'Gauss elimination with backward substitution' (see Project 7).

5.4 Exercises

Obligatory

1. Derive the formula 5.3.5.
2. The power expansion of expression 5.2.6 leads to an approximation (for small h) $\varphi(h) = -0.5h$ (verify). Use this in the BVP1D code (Boundary Value Problem in 1D) to set the second point for the scheme 5.3.5. Evaluate the function $\varphi(r)$ at small variations of $\varphi(h)$,

say in the range $-0.4h \div -0.6h$ (h should be sufficiently small to assure high accuracy). Observe and interpret the behaviour of the function $\varphi(r)$ in reference to the analytical solution (plot both in the same graph). Find the coefficient at h for which both solutions coincide. Is the coefficient the same when h is changed?

3. Modify the code so that the Numerov scheme works 'backwards', that is starting the recursive procedure in 'infinity' (in practice sufficiently far, set the appropriate initial conditions). Compare the resulting function with the analytical one, and draw conclusions.

4. Extend the code so that it calculates the gravitational field from the formula $a_g = -d\phi(r)/dr$. Assuming that the radius of a star is defined as the distance from the centre to the point at which the force reaches maximum (this is also what happens in a planet, and how the radius of an atom is defined), find the radius of the star.

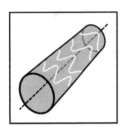

Project 6
Normal Modes in a Cylindrical Waveguide

The eigenvalue problem (EVP) appears, for example, in the description of standing waves or stationary quantum systems. Many theoretical techniques have been developed to solve this problem. Here, a numerical method based on recursive schemes for solving ordinary differential equations is presented: the shooting method. The method, although seemingly simple, since it deals with 1D case and thus can be applied only to low dimensional or high symmetry structures, is employed in real-life research, for example in finding the atomic structure within Density Functional Theory. In this project, a simple classical system is analysed, namely a cylindrical waveguide (e.g. optical fibre) in which the normal modes of a scalar wave have to be found.

6.1 Physics Background: Wave Equation, Standing Waves

Let us begin with recalling the wave equation in 3D, for a scalar wave $\Phi(\mathbf{r}, t)$

$$\nabla^2 \phi - \frac{1}{v_p^2} \frac{\partial^2 \phi}{\partial t^2} = 0, \tag{6.1.1}$$

where $v_p = \omega/k$ is the phase speed, $k = (2\pi)/\lambda$ is the wave number, and $\omega = (2\pi)/T$ represents angular frequency.

We consider the case of axial symmetry (like in the cylindrical waveguide) and write down the Laplace operator in cylindrical coordinates (see Figure 6.1)

$$\nabla^2 \phi = \left(\underbrace{\frac{\partial^2 \phi}{\partial r^2} + \frac{1}{r} \frac{\partial \phi}{\partial r}}_{\frac{1}{r} \frac{\partial \phi}{\partial r} \left(r \frac{\partial \phi}{\partial r} \right)} + \frac{1}{r^2} \frac{\partial^2 \phi}{\partial \varphi^2} + \frac{\partial^2 \phi}{\partial z^2} \right).$$

Due to axial symmetry all the partial derivatives with respect to φ and z disappear, and the wave equation takes the form

$$\frac{\partial^2 \phi}{\partial r^2} + \frac{1}{r}\frac{\partial \phi}{\partial r} - \frac{1}{v_p^2}\frac{\partial^2 \phi}{\partial t^2} = 0. \tag{6.1.2}$$

The field of the scalar waves still remains a function of radial coordinates and time $\phi(r,t)$. In the case of a cylindrical waveguide there are boundary conditions: $\phi(r = 1,t) = 0$, $\phi(r = 0,t) = 1$, $(\partial\phi(r,t)/\partial r)|_{r=0} = 0$, which make the expected solution to be a standing wave along r (we assume the radius of the waveguide $R = 1$ and use the fact the wave can be arbitrarily normalised, that is its amplitude is equal to 1 at the central axis). In such situation the wave function can be represented as the product of only the r-dependent part and only the time-dependent part $\phi(r,t) = \phi_r(r)\phi_t(t)$. When substituting this product into Eq. 6.1.2 and after simple rearrangement we get

$$\frac{1}{k^2}\frac{1}{\phi_r(r)}\left[\frac{d^2\phi_r(r)}{dr^2} + \frac{1}{r}\frac{d\phi_r(r)}{dr}\right] = \frac{1}{\omega^2}\frac{1}{\phi_t(t)}\frac{d^2\phi_t(t)}{dt^2}. \tag{6.1.3}$$

In this equation we have also separated the two constants that comprise the phase speed, ω being the time domain characteristics of a wave, and k the position space characteristics. In this way, on the left-hand side we only have the r-dependent function and on the right side the t-dependent one. It means that both sides must be equal to a constant, and the fact that locally we can expect harmonic oscillations indicates the value -1 for this constant (then, with the time-dependent part it forms an equation for a harmonic oscillator).

The reasoning described here leads to the stationary wave equation

$$\left(\frac{d^2}{dr^2} + \frac{1}{r}\frac{d}{dr}\right)\phi(r) = -(n(r)k)^2\phi(r), \tag{6.1.4}$$

with the boundary conditions $\phi(r=1)=0$, $\phi(r=0)=1$, $(d\phi(r)/dr)|_{r=0} = 0$; additionally, it has been taken into account that in a medium (such as an optical fibre) the wave number can be locally modified by a refraction coefficient $n(r)$, $k'(r) = n(r)k$.

6.2 Problem: Normal Modes in an Optical Fibre

The stationary wave equation derived in previous section is an EVP, thus we expect the solutions in the form of a sequence of pairs $\{(\phi_n(r), k_n)\}$, the so-called normal modes and associated wave numbers k_n. A standard substitution $\phi(r) = \varphi/\sqrt{r}$ allows to simplify the equation to a form suitable for application of the Numerov–Cowell's algorithm

$$\left[\frac{d^2}{dr^2} + \left(\frac{1}{4r^2} + (n(r)k)^2\right)\right]\varphi(r) = 0. \tag{6.2.1}$$

Figure 6.1 Cylindrical coordinates

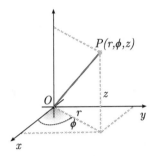

Note that the asymptotic behaviour of the function $\varphi(r)$ in the limit $r \mapsto 0$ must be $\sim \sqrt{r}$ (see the third boundary condition). This fact will be used to avoid singularity in the term $1/(4r)^2$ in numerical calculations. We will solve the problem numerically using the 'shooting method' described next for the EVP.

6.3 Numerical Method: Shooting Method

The idea of the shooting method is simple. We perform the recursive procedure (like the Numerow–Cowell's algorithm) with a certain trial value of k^2, starting at one end of the independent variable interval. When the other end is reached we check the value of the function. If it obeys the boundary condition, then the trial value of k^2 is the eigenvalue. In practice, the problem reduces to finding the roots of the function $F(k^2) = \varphi_{r=1}(k^2)$.

In the case of a 1D quantum well we also deal with the EVP (see also Project 1)

$$\left[-\frac{1}{2}\frac{d^2}{dx^2} + V(x) \right] \psi(x) = \varepsilon \psi(x). \tag{6.3.1}$$

However, the problem here is slightly more complicated. When integrating into the classically forbidden region (negative kinetic energy), if the eigenenergy differs from the true one (even by a very small amount), a rapidly (exponentially) increasing solution appears, and it is very difficult to match the eigenvalue. In this case, the recursion procedure must be conducted from both ends of the domain, that is always from within the classically forbidden region into the allowed one. The criterion for finding the eigenvalue is a smooth connection of the two solutions (the one 'from the left' and 'from the right') at a certain testing point which should be in the classically allowed region (note that the same condition has been used in Project 1, see Figure 1.2). Since the quantum states can formally be arbitrarily normalised, so that

the two solutions connect each other at the testing point, the next condition must be applied, that is their derivatives are equal, which using
the two-point scheme for the derivative gives

$$\frac{\psi_<(x_m - h) - \psi_<(x_m)}{h} = \frac{\psi_>(x_m - h) - \psi_>(x_m)}{h}. \qquad (6.3.2)$$

We reduce h from Eq. 6.3.2 and divide it by $\psi_>(x_m)$ (or $\psi_<(x_m)$)
to have the expressions on both sides typically equal to 1. A small
rearrangement leads to the criterion

$$\frac{1}{\psi_<(x_m)} [\psi_<(x_m - h) - \psi_>(x_m - h)] = 0, \qquad (6.3.3)$$

where $\varphi_<$ and $\varphi_>$ are recursively found functions 'from the left' and
'from the right', respectively, and x_m is the testing point.

Unlike in the case of method described in Project1 (for rectangular
quantum well), this method can be used to solve the EVP for a quantum
well of arbitrary shape, provided it does not contain potential barriers
inside, and it will be used in one of advanced projects (see Project 12).

6.4 Exercises

Obligatory

1. The application of the shooting method in the problem considered
   here involves certain complication, namely although we know the
   initial value of the function $\varphi(r = 0) = 0$, the multiplicative term in
   the stationary wave equation is singular, due to the $1/(4r^2)$ component. To omit this singularity, we have to start the recursive scheme
   at a certain small distance from zero, $r = \varepsilon$. Thus, in principle, two
   control parameters emerge: ε and h (the grid parameter). However,
   to simplify the computation, we will use only one parameter, h, and
   start the recursive scheme at $r = h$. In such an approach it is sufficient
   to study the convergence of results with respect to the grid parameter
   only. Another issue is how to set the values of the function at $r = h$
   and $r = 2h$ (two initial values are needed for the three-point scheme).
   For that purpose we will use the fact that the asymptotic behaviour
   of the function $\varphi(r)$ is known, $\varphi(r \to 0) \to \sqrt{r}$ (see the discussion
   of the problem). Since the function can be arbitrarily normalised, we
   can set the values $\varphi(h) = a\sqrt{h}$ and $\varphi(h) = a\sqrt{2h}$, where a is an
   arbitrary coefficient.
   Run the WAVEGUIDE code and test the convergence of the results
   with respect to the grid parameter h.

2. Calculate the eigenvalues of the wave number and compare the numerical results with the analytical values of 2.404826, 5.520078, 8.653728, and 11.791534. Is it possible to get the exact analytical values from the numerical calculations? (Hint: Note that there are two parameters which affect the results: the grid h in real space, and the uncertainty in the finding roots procedure.)

3. Extend the code so that it outputs the radial functions of normal modes amplitudes. Visualise the results.

Supplementary

1. Modify the code so that it calculates the eigenenergies and eigen-functions for an infinite quantum well and/or the normal modes (functions and wave numbers) for a string fixed at both ends. Compare the results with the well-known analytical values.

2. For the string from previous exercise consider a situation of a non-uniform mass distribution (e.g. linear) and check how it affects the results.

Challenge

1. Consider the case of a non-uniform but still cylindrically symmetric refractive index $k(r) = n(r)k_c$, where k_c is a wave number in a vacuum and $n(r)$ is the radial distribution of the refractive index (must not exceed 2.0 and its value at the wall of the waveguide should be 1.0 – vacuum). How do various functions $n(r)$ affect the wave numbers of normal modes? Try parabolic or exponential functions.

2. Modify the code so that it finds the energy levels of a rectangular quantum well discussed in Project 1, compare the results with the results obtained with the QWELL code.

Project 7
Thermal Insulation Properties of a Wall

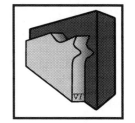

This is the first project devoted to partial differential equations (PDEs), and although, due to symmetry, the problem is still reduced to 1D, and the differential equation becomes ordinary, a numerical technique typically applied to PDEs, finite difference (FD), will be introduced and discussed.

We will study the insulating properties of a house wall, in particular, how temperature across the wall depends on the thermal conductivity distribution of an insulating material. To do this we have to apply the steady-state diffusion equation.

7.1 Physics Background: Steady-State Diffusion

To derive the steady-state diffusion equation we will use two fundamental laws: Fick's law (diffusion) and Continuity law (Figure 7.1), which are expressed, respectively, by the following equations:

$$\mathbf{J}(\mathbf{r}, t) = -D(\mathbf{r}, t)\nabla\phi(\mathbf{r}, t), \tag{7.1.1}$$

$$\nabla \cdot \mathbf{J}(\mathbf{r}, t) = -\frac{\partial\phi(\mathbf{r}, t)}{\partial t} + S(\mathbf{r}, t), \tag{7.1.2}$$

where $\phi(\mathbf{r}, t)$ is the density of the diffusing substance, $\mathbf{J}(\mathbf{r}, t)$ is its flux (current density), $D(\mathbf{r}, t)$ is the diffusion coefficient, and $S(\mathbf{r}, t)$ is the 'source' function, that is the rate at which the density of the substance $\phi(\mathbf{r}, t)$ is produced at the point $\mathbf{r}$.

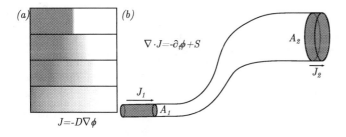

Figure 7.1 Diffusion's laws: (a) Fick's law (b) Continuity law

Figure 7.2 Examples of gradients (arrows) of scalar functions represented by a colour intensity scale

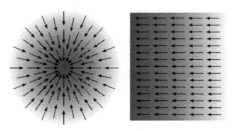

The 'nabla' vector differential operator reads

$$\nabla = \widehat{i}\frac{\partial}{\partial x} + \widehat{j}\frac{\partial}{\partial y} + \widehat{k}\frac{\partial}{\partial z}, \qquad (7.1.3)$$

and plays the role of a gradient (Figure 7.2) in the first equation, and a divergence in the second one.

Note that all the quantities appearing in the equations till now are, in general, functions of position and time, but in the next step we assume that there are no variations with time (stationary system), and substitute Eq. 7.1.1 into Eq. 7.1.2. As a result we get

$$\nabla \cdot [D(\mathbf{r})\nabla\phi(\mathbf{r})] = -S(\mathbf{r}). \qquad (7.1.4)$$

The $S(\mathbf{r})$ and $\phi(\mathbf{r})$ may have different meanings, depending on the type of system considered. In particular, they can be interpreted as the source of heat and temperature, respectively. In that case, the diffusion coefficient $D(\mathbf{r})$ represents the thermal conductivity.

The equation is an elliptic partial differential equation which, together with the boundary conditions (values of the function and/or values of its derivatives at some borders surrounding the region of interest, and sometimes also inside the region), forms the boundary value problem. It is interesting and surprising to note that at a constant diffusion coefficient Eq. 7.1.4 becomes the Poisson's equation.

7.2 Problem: Steady-State Diffusion of Heat through the Wall

In this project, we will consider the quasi-1D case (such a statement means that although the system is three-dimensional, all the quantities vary only with one spacial variable). This is the heat diffusion through a wall, say along the X axis, and then the equation simplifies to

$$-S(x) = \frac{d}{dx}\left(D(x)\frac{d\phi(x)}{dx}\right), \qquad (7.2.1)$$

or

$$-S(x) = D'(x)\frac{d\phi(x)}{dx} + D(x)\frac{d^2\phi(x)}{dx^2}, \qquad (7.2.2)$$

with known values of the function ϕ at the interval ends: $\phi(x = 0) = \phi_0$ and $\phi(x = L) = \phi_N$ (where L is the wall thickness).

We assume that the wall is in contact with the heat reservoirs at both sides, which stabilises the temperature, so that ϕ_0 and ϕ_N correspond to the temperature at both ends. Moreover, we assume that there are no heat sources inside the wall ($S(x) = 0$), which means that the temperature distribution across the wall must be monotonic, no matter the heat conductivity.

7.3 Numerical Method: Finite Difference Method

In this project, the FD method will be applied. First, a regular grid is introduced, with a grid parameter h. Second, the differential equation is discretised, using the three-point formula for derivatives. In the example considered, the discretization of Eq. 7.2.2 takes the form

$$-S_i = D_i'\frac{\phi_{i+1} - \phi_{i-1}}{2h} + D_i\frac{\phi_{i+1} + \phi_{i-1} - 2\phi_i}{h^2}, \qquad (7.3.1)$$

or

$$\left(D_i - \frac{h}{2}D_i'\right)\phi_{i-1} - 2D_i\phi_i + \left(D_i + \frac{h}{2}D_i'\right)\phi_{i+1} = -S_ih^2. \quad (7.3.2)$$

As a result we obtain a system of linear equations where the values of the function ϕ at grid nodes are unknown. Thus, the FD method converts the boundary value problem for differential equations into a system of linear equations.

In our case this is a tri-diagonal system of linear equations of the form

$$\left\{ A_i^-\phi_{i-1} + A_i^0\phi_i + A_i^+\phi_{i+1} = b_i. \right. \qquad (7.3.3)$$

Note that the boundary conditions ϕ_0 and ϕ_N have to be moved to the right-hand side of the equation.

To solve this system of equations we will use the very efficient Gaussian elimination with back-substitution algorithm. We assume that the solution satisfies a one-step forward linear recursion relation of the form

$$\phi_{i+1} = \alpha_i \phi_i + \beta_i, \qquad (7.3.4)$$

where α_i and β_i are the coefficients to be determined. Substituting this into Eq. (7.3.3), we get

$$A_i^- (\alpha_i \phi_i + \beta_i) + A_i^0 \phi_i + A_i^+ \phi_{i-1} = b_i, \qquad (7.3.5)$$

which can be solved for ϕ_i to yield

$$\phi_i = \gamma_i A_i^- \phi_{i-1} + \gamma_i (A_i^+ \beta_i - b_i), \qquad (7.3.6)$$

with

$$\gamma_i = -\frac{1}{A_i^0 + A_i^+ \alpha_i}. \qquad (7.3.7)$$

Upon comparing Eq. (7.3.4) with Eq. (7.3.6) we obtain the back-ward recursion relations for the $\alpha's$ and $\beta's$

$$\alpha_{i-1} = \gamma_i A_i^-, \qquad (7.3.8)$$

$$\beta_{i-1} = \gamma_i (A_i^+ \beta_i - b_i). \qquad (7.3.9)$$

The strategy now is as follows: we use the recursion relations (7.3.8, 7.3.9) to determine the α_i and β_i, for i running from $N-2$ down to zeros. The starting values to be used are

$$\alpha_{N-1} = 0, \beta_{N-1} = \phi_N, \qquad (7.3.10)$$

which guarantee the correct value of ϕ at the upper boundary. Know-ing these coefficients, we use the recursion relation (7.3.4) in a forward sweep from $i = 0$ to $N - 1$ to find the solution, with the starting value ϕ_0 known from the boundary condition. Thus, the solution is deter-mined in only two sweeps of the involving arithmetic operations of the order N.

Note that this technique can be also applied to the boundary value problem discussed in Project 5 (Gravitation inside a star). The method of solving the tri-diagonal system of linear equations can be treated here as a tricky way of 'moving' one of boundary conditions to the opposite side.

At the end we should refer to the case of partial differential equation in many dimensions, in the context of the FD method. We will do it on the example of the Laplace elliptic partial differential equation

$$\left(\frac{\partial^2}{\partial x^2} + \frac{\partial^2}{\partial y^2} + \frac{\partial^2}{\partial z^2} \right) \phi(x, y, z) = 0, \qquad (7.3.11)$$

with the boundary conditions in the form of values of ϕ at a boundary and some places inside the region of interest (Dirichlet type), and/or derivatives of ϕ at the boundary, in normal directions (von Neuman type).

We limit the discussion to 2D to shorten the equations, but the methods presented also apply to 3D cases. Thus, in the first step, the differential equation is discretised by introducing a regular grid in the domain at a given coordinate system, that is the differential operators are converted into the FD operators, using the three-point formula

$$\frac{\phi_{i+1,j} + \phi_{i-1,j} - 2\phi_{i,j}}{h_x^2} + \frac{\phi_{i,j+1} + \phi_{i,j-1} - 2\phi_{i,j}}{h_y^2} = 0, \qquad (7.3.12)$$

where the labels $i = 1, \ldots, N_x, j = 1, \ldots, N_y$ indicate the grid points along x and y, respectively, with N_x, N_y being the respective numbers of the grid points. Next, we have to create a mutually unambiguous mapping of the double indexes onto single ones, for example $(i,j) \rightarrow k = i + N_y(j - 1)$. Now, the set of values $\{\phi_{i,j}\}$ becomes a vector $\{\phi_k\}$, and Eq. 7.3.12 clearly appears as the system of linear equations of the form

$$\hat{A}\phi = a, \qquad (7.3.13)$$

where the vector a is formed from the values of the ϕ function at the grid points at which the values are known from the boundary conditions. Thus, the problem is reduced to the inversion of the $\hat{A}$ matrix

$$\phi = \hat{A}^{-1}a, \qquad (7.3.14)$$

which seems not be a trivial operation considering the fact that the matrix can be very big. For example, in 3D, at the grid $100 \times 100 \times 100$ (which is not particularly dense) the order of the matrix is 10^6. However, a good news is that this is a so-called sparse matrix in which a vast majority of elements are zero, and there are special methods of their storage and very effective algorithms of their inversion. Finally, it should be mentioned that a great shortcoming of the FD methods is that the shape of the considered domain as well as the related regular grid must fit the chosen coordinate system. This limitation does not appear in the finite elements methods, which are discussed in the next project.

Figure 7.3 The considered step function as a model of a wall with insulation

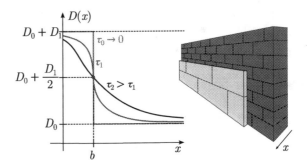

7.4 Exercises

Obligatory

1. Using the FTABLE code tabulate a few chosen functions of the thermal conductivity coefficient $D(x)$ and their first derivatives $D'(x)$ (e.g. linear, parabolic, nonmonotonic – such as sinefunction, etc.). In particular, consider a step function (Fermi–Dirac type) which would emulate the wall consisting of two layers: a high thermal conductivity material (mechanically strong), and a low thermal conductivity material (mechanically weak), for example a brick and a styrofoam (Figure 7.3). The Fermi–Dirac (step-like) function is expressed by

$$D(x) = \frac{D_1}{1 + exp(\frac{x-b}{\tau})} + D_0, \qquad (7.4.1)$$

where the τ parameter is responsible for the step smearing (for small τ the step is sharp), b denotes a coordinate of the border between the two materials, $D_1 + D_0$ and D_0 are thermal conductivities of higher and lower conductivity materials, respectively. Evaluate the derivative of the function and tabulate both $D(x)$ and $D'(x)$.

2. Test the DIFFUSION code for a uniform linear equation $\frac{d^2y}{dx^2} = 0$ ($D' = 0, D = 1$) with different boundary values and grid parameters.

3. Calculate the temperature across the wall for thermal conductivities tabulated in the first exercise. Is the temperature distribution always monotonic, even if the thermal conductivity is non-monotonic? Check what is the effect of reflecting the thermal conductivity function with respect of the central point? Does the speed of the flow change? Is the temperature distribution across the wall for the step-like thermal conductivity as you expected? What is better from a

practical point of view, to put the styrofoam layer inside or outside the building? Why?

Challenge

1. Introduce 'heat sources' inside the wall by setting additional conditions in the form of the values of $\phi(x)$ function at chosen points (there can be more than one point) (Figure 7.3). Observe the effect of the additional sources on the temperature distribution and on the heat flow (see Figure 7.4).

Figure 7.4 The considered source of heat can be heating pipes inside a well, whose temperature would be constant

Project 8
Cylindrical Capacitor

The electrostatic problems described by Poisson's equation, that is finding the electric potential for given boundary conditions and charge
density distribution, have their variational formulation (as many other
problems in physics). In physics language this principle says that the
potential being the solution minimises the total energy of the system
(energy of the field plus the potential energy of the charge). However, there also exists a rigorous mathematical proof that solving the
Poisson's differential equation is equivalent to the problem of minimisation of a certain functional (the variational principle), thus the
method can be extended on the systems where the 'energy' interpretation does not apply. The variational principles constitute a base
on which a very important and widely employed class of numerical
techniques is based – the finite element (FE) methods, which will
be discussed in this project. A great advantage of these methods is
a big freedom in the choice of points (the grid) in the independent
variable space, which is not limited by the choice of the coordinate
system, and also can be adjusted to the expected behaviour of the
function we want to find (denser grid in the regions of expected big
variations of the function). For the sake of simplicity, a high symmetry (cylindrical) electrostatic system has been chosen which can
be described by 1D equation. This allows to present the ideas while
significantly simplifying the computational procedure. However, in
Project 14 the discussion will be extended to 2D (which also applies
to 3D).

The system considered here is a capacitor consisting of two coaxial metallic cylinders of different radii. This is not a purely academic
problem since an important practical reference of such a system exists,
namely the coaxial transmission cables (antenna, telecommunication,
etc.) whose main specification is the capacitance per unit of length,
which determines the cable frequency-dependent transmittance and
thus its transmission spectrum (one of the exercises is devoted to the
calculation of capacitance).

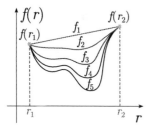

Figure 8.1 The variational principle tests a series of functions to minimise functional

8.1 Physics Background: Variational Principle for Electrostatic Systems

An electrostatic system is described by Poison's equation (see Project 5)

$$- \nabla^2 \phi(\mathbf{r}) = 4\pi \rho(\mathbf{r}), \qquad (8.1.1)$$

which, for a given charge density distribution $\rho(\mathbf{r})$ and boundary values, forms the boundary value problem for an elliptic partial differential equation (see Project 5), the solution to which is the electric potential function $\phi(\mathbf{r})$ (Figure 8.1). However, an alternative approach to the problem exists, the *variational principle*. It can be formally proved (Chapter A.5) that the function $\phi(\mathbf{r})$ being the solution to Eq. (8.1.1) minimises the functional

$$F[\phi] = \int d^3\mathbf{r} \left[\frac{1}{2}(\nabla\phi(\mathbf{r}))^2 - 4\pi\rho(\mathbf{r})\phi(\mathbf{r}) \right]. \qquad (8.1.2)$$

In the case of an electrostatic system the functional above corresponds to the total energy, but since the proof is general it can be treated just as a generic functional in other cases. The problem is then converted into the problem of finding the minimum of Eq. (8.1.2). Upon any parametrisation of the function $\phi(\mathbf{r})$ with a set of parameters $\{\alpha_i\}$, the functional becomes a function $F(\{\alpha_i\})$ and the problem reduces to finding the minimum of the function. In the *finite elements* (Figure 8.2)

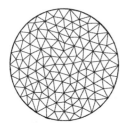

Figure 8.2 The finite elements method applied to a circle

method, the function is represented by its values at the grid nodes $\{\phi_i\}$, which become the parameters to be found. At variance with the *finite difference* method the grid can be shaped in an arbitrary way (independent of the choice of the coordinate system); it can be denser in the regions where the function is expected to vary at a high rate, or can have a triangular rather than rectangular geometry.

8.2 Problem: Cylindrical Capacitor

The system considered in this project has cylindrical symmetry. The Poisson's equation takes the form (see Eq. 6.1, and the preceding discussion)

$$-4\pi\rho(r)r = \frac{\partial}{\partial r}\left(r\frac{\partial\phi}{\partial r}\right),\qquad(8.2.1)$$

which together with the boundary values $\phi(r_1)$ and $\phi(r_2)$ leads to the boundary value problem (similar to that discussed in Project 5). Equation (8.2.1) is also formally equivalent to the diffusion equation discussed in the previous project, and thus the method introduced there can be also applied. In this project, however, we will use the variational principle. A respective functional in the cylindrical coordinates has the form

$$F[\phi] = \int_{r_1}^{r_2} dr \left[\frac{1}{2}\left(\frac{d\phi(r)}{dr}\right)^2 - 4\pi\rho(r)\phi(r)\right]r.\qquad(8.2.2)$$

8.3 Numerical Method: Finite Elements (FE) Method

The discrete representation of the functional 8.2.2 will be based on the simplest possible numerical quadrature scheme, the rectangles method

$$\int_{r_1}^{r_2} drf(r) \approx h \sum_{i}^{N} \frac{1}{2}(f_i + f_{i-1}),\qquad(8.3.1)$$

where $\frac{1}{2}(f_i + f_{i-1})$ is the value of the function $f(r)$ in the middle of the interval (r_i, r_{i-1}) at a linear approximation of the function within the interval.

Equation (8.3.1), when applied to 8.2.2 leads to (do the derivation)

$$F[\{\phi_i\}] \approx \frac{1}{2h}\sum_{i}^{N}(\phi_i - \phi_{i-1})^2 r_{i-\frac{1}{2}} - 2\pi h \sum_{i}^{N}(\phi_i\rho_i r_i + \phi_{i-1}\rho_{i-1}r_{i-1}),$$

$$(8.3.2)$$

where $r_i = r_1 + ih$, $r_{i-1/2} = r_i - \frac{1}{2}h$, $h = (r_2 - r_1)/N$ and F becomes a function of many variables $F[\phi] \rightarrow F(\{\phi_i\})$.

The condition for the minimum

$$\frac{\partial F}{\partial \phi_i} = 0 \tag{8.3.3}$$

leads to

$$\phi_{i+1}r_{i+\frac{1}{2}} - \phi_i(r_{i+\frac{1}{2}} + r_{i-\frac{1}{2}}) + \phi_{i-1}r_{i-\frac{1}{2}} + h^2 4\pi\rho_i r_i = 0. \tag{8.3.4}$$

This is a starting equation for the Gauss–Seidel iterative minimisation procedure. Let us solve it for ϕ_i

$$\phi_i = \phi_{i+1}\frac{r_{i+\frac{1}{2}}}{2r_i} + \phi_{i-1}\frac{r_{i-\frac{1}{2}}}{2r_i} + 2\pi h^2 \rho_i, \tag{8.3.5}$$

or

$$\phi_i = \frac{1}{2}(\phi_{i+1} + \phi_{i-1}) + \frac{h}{4r_i}(\phi_{i+1} - \phi_{i-1}) + 2\pi h^2 \rho_i. \tag{8.3.6}$$

The strategy is then as follows:

i. Make an initial guess of the function ϕ (e.g. the linear function between $\phi_0 = \phi(r_1)$ and $\phi_N = \phi(r_2)$).

ii. Calculate the new value of ϕ_i form Eq. (8.3.6) at each point of the grid r_i (except for the boundary values); we will call this step a 'sweep'.

iii. At the new set $\{\phi_i\}$ (after the sweep) calculate the value of the functional (8.3.2).

iv. Repeat the whole procedure until Eq. (8.3.2) stabilises, that is the criterion that in three subsequent sweeps the functional does not vary more than given ϵ is fulfilled.

This procedure can be further improved by mixing the result of a sweep with the previous one. This is done via the mixing parameter ω in the following way:

$$\phi_i' = \phi_i^{old}(1 - \omega) + \phi_i^{new}\omega, \tag{8.3.7}$$

where ϕ_i^{new} is the result of the last sweep, ϕ_i^{old} is the result of the previous sweep, and ϕ_i' is the value to be taken in the nest sweep.

It can be formally proven that the procedure is convergent for a certain range of ω values, but here we will find the range by performing numerical experiments (as one of the exercises). In all the exercises (except one) we will also assume that the density of the charge ρ is zero in the space between metallic cylinders; only the values of the potential at the cylinders $\phi(r_1)$ and $\phi(r_2)$ are given (boundary conditions). This means that we are in fact solving Laplace's equation.

8.4 Exercises

Obligatory

1. Test the CAPACITOR code for different boundary conditions and control parameters. Check the convergence of the functional F with respect to the grid parameter, and find the optimum value of h, that is that value at which the energy assumes minimum. Note that at too small values of h the energy increases. The reason for this can be found in Eq. 8.3.2, where the term containing the difference between neighbouring potentials appears (tending to zero when h is too small). Similar situation takes place in the Gauss–Seidel formula 8.3.6, thus the last exercise in this list must be performed with caution as for the choice of h.

2. Investigate into the effect of 'mixing'. Find the range of the parameter ω at which the procedure is convergent and find the value at which the relaxation is the fastest. (Hint: Modify the code so that it outputs the values of the F functional as a function of the iteration number, and plot the functions for different computational parameters.)

3. Modify the code so that it takes into account a non-zero density distribution in Eq. (8.3.6) and repeat the above test for some model density distribution $\rho(r)$ of your own choice. Find the optimum value of ω. Repeat it for two to three different model densities.

4. Perform fully converged calculations for chosen boundary conditions and compare the results with the analytical solution $\varphi(r) = A \ln(r) + B$ (the values of constants A, B should be found from boundary conditions, by solving a system of two linear equations).

Advanced Projects

The following projects are advanced, which means they involve more complex physical systems. Both the physics background and numerical methods required do not exceed the material covered in the first eight projects, so no theoretical introductions are provided. The primary formulations of the presented problems have been solved, and their respective computational codes can be found in an online repository. However, undertaking these projects offers ample opportunity for individual creativity, posing new problems, asking questions, and discovering answers. These projects are specifically designed for students with a keen interest in research utilising computational methods.

Project 9
Coupled Harmonic Oscillators

The first advanced project refers to Project 3, where numerical methods for solving differential equation were presented. Every physicist in their career realises that a model of a harmonic oscillator is relevant in numerous fields of science, including mechanics, atomic and quantum physics, optics or electronics. It is frequently used to describe physical phenomena present in natural and artificial systems. Here, we propose to expand this problem to a system of coupled oscillators. Coupled oscillators are relevant in both nature and technology and can be a useful model to describe the physics of musical instruments, electronic devices used for communication, electromagnetic wave and matter interactions or biological systems such as the function of a heart and its irregularities. Coupled oscillators have been studied by Lionel Robert Wilberforce in a laboratory setting, where he built the famous pendulum named after him. The Wilberforce pendulum consists of a weight suspended from a spring which twists as the spring stretches and retracts. During the experiment, Wilberforce observed the energy transfer between two degrees of freedom of oscillations: rotation around the spring axis and longitudinal movement along that axis, while the total energy was conserved (except for the losses due to friction).

9.1 Problem: Equations of Motion of Coupled Oscillators

Coupled oscillators can be mathematically expressed through equations of motion of each of the oscillators, with the coupling term (their interaction) taken into account. As an example we can consider two simple cases (Figure 9.1). The first one is a 1D system of two point masses suspended from two springs of k_1 and k_2 constants, connected with a third spring of k_3 spring constant. The second one is a system of two simple pendulums moving in the same plane, where the masses are connected with a spring. In addition, in Ex. 4 the Wilberforce pendulum will be studied, as it is an example of a coupled oscillator as well.

For simplicity, we assume that both masses are the same and $m = 1$, which means that their frequencies $\omega = \sqrt{k/m}$ depend only on the constant k. In the first case, the equations of motions are expressed as

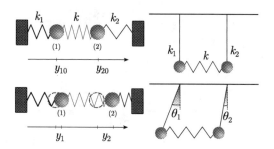

Figure 9.1 Two possible realisations of a coupled harmonic oscillator

$$d^2y_1/dx^2 = -k_1y_1 - k(y_1 - y_2),$$
$$d^2y_2/dx^2 = -k_2y_2 - k(y_2 - y_1),$$

(9.1.1)

where y is the position of the mass relative to the equilibrium. In the second case (refer to Project 3, Figure 3.1), we know that in a simple pendulum the driving force is the projection of the gravitational force on the tangent of the arc of motion. The equation of motion is then

$$\frac{d^2s}{dt^2} = -g\sin\theta,$$

(9.1.2)

where s is the position of the point on the arc. For small angles (which is what we consider here) we can assume that s is related to y by the relation $y = l\sin(\theta)$ on the horizontal axis. The equation of motion can then be written as

$$\frac{d^2y}{dt^2} = -\frac{g}{l}y.$$

(9.1.3)

It is a slightly modified form of Eq. 3.2.3 from Project 3, where instead of y we had $\theta = s/l \approx y/l$. However, this form allows to introduce coupling between the two pendulums: a spring connecting the two masses:

$$d^2y_1/dx^2 = -\frac{g}{l_1}y_1 - \frac{k}{m_1}(y_1 - y_2),$$
$$d^2y_2/dx^2 = -\frac{g}{l_2}y_2 - \frac{k}{m_1}(y_1 - y_2).$$

(9.1.4)

If we assume that the masses are the same and equal to 1, $m_1 = m_2 = 1$ (which means that the frequency of the oscillations is controlled only with the length of the pendulum), and we use the relations $g/l_1 = k_1$ and $g/l_2 = k_2$, we can arrive at a system of equation identical to Eq. 9.1.1. This means that the two qualitatively different systems can be described by the same mathematical model. What is left is to take a closer look at the nature of solving those equations, and through computational experimentation getting familiar with the physics of this system, its behaviour depending on the parameters of the pendulums and their coupling, and, in particular, the energy transfer between the pendulums. A system of coupled oscillators, described with Eq. 9.1.1, is an IVP for ordinary differential equations. The numerical methods necessary to solve that problem have already been discussed in Project 3. In this project, the Runge–Kutta algorithm will be used, as it has been identified as the most effective in previous Project.

9.2 Exercises

Compound Pendulum

1. In the program CHO, set the coupling constant $k = 0$ and check the convergence of results with respect to the time-step h. Observe the total energy of the system which should be conserved during simulation. Repeat this exercise for different spring constants k_1 and k_2. Check whether the frequencies of oscillators agree with the analytical values.

2. Perform simulations of two simple cases: *Sloshing mode* and *Breathing mode*, see Figure 9.2. In order to do this, set appropriate initial conditions, and then find the dependence on time of the total energy of the system and energies of both oscillators. Interpret the results. Check how the coupling constant k affects such oscillations.

3. The most interesting phenomenon appearing in coupled oscillators is the energy flow between them. This phenomenon finds many references in nature and technology. For example, in physics of musical instruments – in violin and similar instruments – the string oscillations are transferred to oscillations of wooden plates and the air cavity of resonant box. In electronics, RLC systems, coupled via electromagnetic waves, can transfer energy between themselves (e.g.

Figure 9.2 Special cases of coupled oscillator's behaviour: (a) sloshing mode, (b) breathing mode

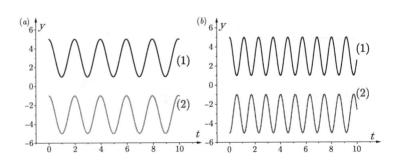

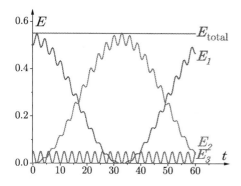

in radiocommunication, telecommunication, contactless control systems, and electronic bank cards). In quantum physics, when an energetically excited quantum system (e.g. in an atom) falls down to the ground state, it can emit a photon which, in turn, can excite another quantum system. These are only examples of phenomena for which coupled oscillators serve as a basic classical model. It should be pointed out that the phenomenon of resonance plays an important role here. Namely, the energy transfer is most effective if it is of a resonant character, that is the agent exciting an oscillator (originating from another oscillator) has the same frequency as its natural frequency, which is called resonant coupling. This exercise gives an opportunity to investigate into the phenomenon.

Resonant Coupling. Consider weakly coupled oscillators ($k \ll k_1, k_2$). For equal spring constants $k_1 = k_2$, find how the frequency of energy flow between oscillators depends on the coupling constant k (Figure 9.3). Try to formulate a universal conclusion by finding the dependence of the frequency on the ratio $k/k_1 = k/k_2$.

Non-resonant Coupling. For weakly coupled oscillators, find the dependence of energy flow between oscillators on the difference between spring constants k_1 and k_2. To do it systematically, try to find the dependence of the ratio of amplitudes E_1^{max}/E_2^{max} on the ratio of spring constants k_1/k_2. How does the coupling constant affect this dependence? Observe the frequency of energy flow and try to find similar dependencies for the frequency.

4. *Wilberforce Pendulum.* Let us consider a system mentioned in the introduction to this project, that is the Wilberforce pendulum. The system consists of a spring of negligible mass, a spring constant k, and torsional spring constant δ. A body of mass m and moment of inertia I with respect to the vertical axis z is suspended from the spring (Figure 9.4). We will assume a linear dependence of

Figure 9.4 The Wilberforce pendulum model with dependencies of displacement z and twist θ angle on time, equivalent to typical dependencies in coupled oscillators

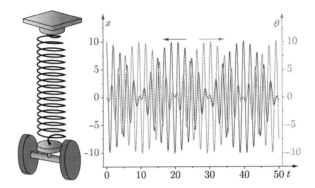

Figure 9.5 Trajectory of a point on the border of the body. Scale of brightness corresponds to time scale (dark colour represents first moments of motion which begins in the point $(0, 1, 10)$, bright – the last moments)

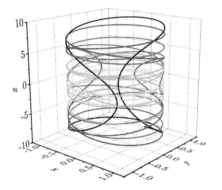

forces (torque) on the linear (angular) displacement, and a bilinear dependence of coupling between the two degrees of freedom $\frac{1}{2}\epsilon z\theta$, where ϵ is a coupling constant. We begin the derivation of the equations of motion from writing down the Lagrangian of the system

$$L = \frac{1}{2}m\dot{z}^2 + \frac{1}{2}I\dot{\theta}^2 - \frac{1}{2}kz^2 - \frac{1}{2}\delta\theta^2 - \frac{1}{2}\epsilon z\theta. \tag{9.2.1}$$

The use of the Euler–Lagrange equations leads to

$$m\ddot{z} = -kz - \frac{1}{2}\epsilon\theta,$$
$$I\ddot{\theta} = -k\theta - \frac{1}{2}\epsilon z.$$

Modify the CHO program so that it simulates the motion of a Wilberforce pendulum, with the possibility of setting different values of parameters. The program should output the values of the total

energy and its components (potential and kinetic energies, of trans-
lational and torsional motions), and position of chosen point in the
body as dependent on time in 3D cartesian coordinates.

Using the modified program simulate the motion of the Wilberforce
pendulum. As a test for correct simulation check the dependence of
the total energy on time (it should be conserved). Plot the dependencies of all the energy components on time. For a chosen point in the
body plot the 3D trajectory of the motion (e.g. Figure 9.5). Repeat
the calculations for a few significantly different sets of parameters.

5. *Lissajous curves.* Jules Antoine Lissajous used two vibrating tuning
forks with mirrors to generate figures which were named *Lissajous
curves*, following his name. To obtain this result using the CHO
code, consider a system of independent oscillators ($k = 0$). In this
case, find the appropriate initial value and draw Lissajous curves in
Cartesian coordinates ($y_1 = x, y_2 = y$) (Figure 9.6a) and polar coordinates ($y_1 = r, y_2 = \theta$) (Figure 9.6b). Next, check what happens
if the oscillators are not independent; consider only a really weak
interaction. Consider the system when $k_1 = k_2$. What is the motion
of the oscillator with and without interaction?

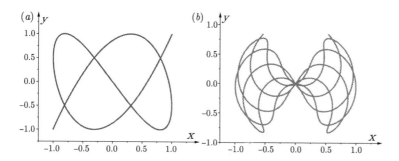

Figure 9.6 The examples of
Lissajous curves: (a) in
Cartesian coordinates, (b) in
Polar coordinates

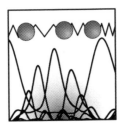

Project 10

The Fermi–Pasta–Ulam–Tsingou Problem

At the beginning of this book we discussed a simple problem of a harmonic oscillator (Project 3), which we then expanded into a problem of coupled oscillators (Project 9). Throughout this project, however, we considered only linear interactions between the oscillators. We also covered methods of solving differential equations and their use in initial value problems (IVP). In this project, we consider a chain of many bodies interacting with each other through a non-linear force. In literature, it is known as the Fermi-Pasta–Ulam–Tsingou problem. The analysis of the physical properties of such a system leads to unexpected observations. First, we see a spontaneous generation of oscillation modes of higher frequencies (i.e. frequency multiplication). Second, from a seemingly chaotic motion an order emerges, which includes a periodicity of that motion in time. This has been possible through the numerical methods described in this book.

10.1 Problem: Dynamics of a One-Dimensional Chain of Interacting Point Masses

Let us consider a chain of interacting point masses m (Figure 10.1). Their equilibrium positions are expressed by a vector of coordinates on the X axis, $x = (\ldots, x^o_{i-1}, x^o_i, x^o_{i+1}, \ldots)$. Let's also assume that the distance between neighbouring masses is constant $x_i - x_{i-1} = a$, and call it the lattice constant a. The masses are allowed to move along the X axis, which is expressed by a time-dependent displacement vector $y =$

Figure 10.1 Chain of oscillating masses

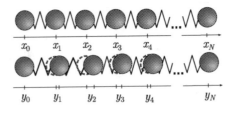

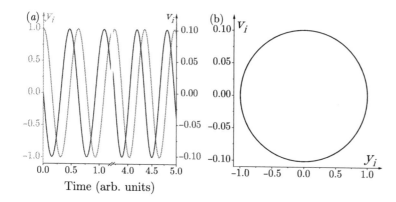

Figure 10.2 Coordinates of the central mass in a harmonic motion $N = 32$, $a = 1$, $m = 1$, $\alpha = 1$, $\beta = 0$: (a) time dependent, (b) in the form of phase diagram

$(\ldots, y_{i-1}, y_i, y_{i+1}, \ldots)$, where $y_i = x_i - x_i^o$ and x_i is the time-dependent position of the mass.

Newton's equations of motion for the i-th mass

$$\frac{\partial^2 y_i}{\partial t^2} = F(y_{i+1} - y_i) - F(y_i - y_{i-1}). \qquad (10.1.1)$$

The expression for the force (right side of the equation) suggests that only the nearest-neighbour interactions are considered, and that the force depends on the difference in the displacements (a relative displacement). Let's start from a linear dependence of the force on the relative displacement $F(y_{i+i} - y_i) = \alpha(y_{i+1} - y_i)$, which simplifies Eq.10.1.1 to

$$\frac{\partial^2 y_i}{\partial t^2} = \alpha(y_{i+1} + y_{i-1} - 2y_i). \qquad (10.1.2)$$

With this form of the equation, we expect that locally the oscillations are harmonic (Figure 10.2), and the motion of two different masses can differ only by a phase factor (due to their periodic symmetry). Taking those two factors into consideration, we can postulate a solution

$$y_i = Ae^{i(\omega t - kx_i)}. \qquad (10.1.3)$$

It is a function that describes a wave, where $\omega = \frac{2\pi}{T}$ is the angular frequency (T is the period of oscillation) and $k = \frac{2\pi}{\lambda}$ is the wavenumber (λ is the wavelength, equal to or greater than the minimum wavelength allowed in the system, $\lambda \geq 2a$). A substitution of Eq. 10.1.3 in Eq. 10.1.2 leads to an expression for the dispersion relation, a detailed derivation of which can be found in the supplementary material (Appendix A.4)

$$\omega(k) = \sqrt{\frac{2\alpha}{m}[1 - \cos(ka)]}, \qquad (10.1.4)$$

with k varying between $-\frac{\pi}{2}$ and $\frac{\pi}{2}$ (first Brillouin Zone). If we limit the length of the chain, assuming that masses x_0 and x_N are immobilised, that is $y_0 = y_N = 0$, then only standing waves (or their superpositions) can be excited in the chain, for which Eq. 10.1.4 can be factorised into spatial temporal components (see Project 6)

$$y = A \cos(\omega t) \sin(kx). \tag{10.1.5}$$

The left boundary condition for the spatial part is automatically fulfilled, due to the choice of the sine function. In order to fulfil the right boundary condition, the following condition must be fulfilled

$$y_N = A \sin(kaN) = A \sin(kL) = 0,$$

where $L = aN$ is the total length of the chain. This leads to a discretisation of allowed wave modes (normal modes)

$$k^n L = n\pi,$$

which means that the wavevector can only have discrete values

$$k^n = \frac{n\pi}{L},$$

with $n = 1, \ldots, N - 1$.

For a given chain of length L, the longest normal mode is always $\lambda_{max} = 2L$, while the shortest is the above-mentioned shortest wavelength $\lambda_{min} = \frac{2L}{N-1}$. This means that the set of allowed modes is not only discrete but also finite, because we can have at most $N - 1$ normal modes. The set can be expressed as

$$\{y^{n0} = A^n \sin(k^n x_1^o), \ldots, \sin(k^n x_{N-1}^o)), \quad n = 1, \ldots, (N - 1)\}. \tag{10.1.6}$$

In the above equation, A^n is the amplitude of the n-th mode and the upper index 0 indicates $t = 0$ in Eq. (10.1.5). The normal modes make up an orthonormal base in $N - 1$ dimensional space, which can be normalised by an appropriate choice of the amplitudes A^n, so that the following relation is fulfilled

$$y^{i0} \cdot y^{j0} = \delta_{ij}.$$

An inclusion of the time-dependent part results in

$$y^n(t) = y^{n0} \cos(\omega^n t), \tag{10.1.7}$$

where $\omega^n = \omega(k^n)$ is the angular frequency associated with the n-th mode, given by the dispersion relation (10.1.4).

We can now derive the vector of velocities in the n-th mode as a time derivative of the displacement vector (10.1.7)

$$v^n(t) = -y^{n0}\omega^n \sin(\omega^n t). \qquad (10.1.8)$$

This allowed us to find the trajectory in a multidimensional phase space for each mode.

Let's include a non-linear, second-order interaction

$$F(y_{i+1} - y_i) = \alpha(y_{i+1} - y_i) + \beta(y_{i+1} - y_i)^2. \qquad (10.1.9)$$

We assume that the initial state of the chain is the first normal mode from the y^{n0} set, with an amplitude A

$$y(t = 0) = \frac{A}{A^1}\bar{y}^{10} = A(\sin(k^1 x_1^o), \ldots, \sin(k^1 x_{N-1}^o)), \qquad (10.1.10)$$

with zero initial velocities $\bar{v}(t = 0) = (0, \ldots, 0)$.

The energy of such a chain is therefore initially only the potential energy of the interaction

$$E = \sum_{i=1}^{N} U(y_i^{lo} - y_{i-1}^{lo}), \qquad (10.1.11)$$

where $U(y) = -(\frac{1}{2}\alpha y^2 + \frac{1}{3}\beta y^3)$, and should be conserved throughout the entire motion.

Therefore, we have a system of differential equations (10.1.1) to solve. We use numerical methods, similarly as in the pendulum (Project 3), and coupled oscillator projects (Project 9). As a result, we obtain a time evolution of the trajectories in a multidimensional phase space of the coordinate and velocity vectors $y(t)$ and $v(t)$. The observation of the time evolution of a trajectory of a chosen mass (which can be done by plotting velocity as a function of position, see Figure 10.3) leads to

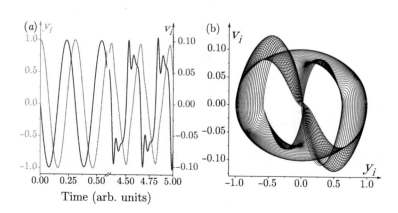

Figure 10.3 Coordinates of the central mass in a non-harmonic motion $N = 32$, $a = 1$, $m = 1$, $\alpha = 1$, $\beta = 0.25$: (a) time dependent, (b) phase diagram

the conclusion that the motion is chaotic. However, a certain order can be observed within that chaos. To do this observation, let us project the time-dependent displacements vector into fundamental modes, which, as we can remember, form an orthonormal basis

$$y(t) = \sum_{n=1}^{N-1} a^n(t)\hat{y}^{n0}. \tag{10.1.12}$$

The coefficients a^n are projections of the state of the system onto the fundamental modes

$$a^n(t) = y(t) \cdot \hat{y}^{n0}. \tag{10.1.13}$$

A similar representation can be performed for the velocity (using the same basis)

$$v(t) = \sum_{n=1}^{N-1} b^n(t)\hat{y}^{n0}, \tag{10.1.14}$$

where the coefficients b^n are

$$b^n(t) = v(t) \cdot \hat{y}^{n0}. \tag{10.1.15}$$

By doing this, we moved the entire time dependence to the coefficients $\{a^n(t)\}$ and $\{b^n(t)\}$. Now, instead of tracking the position and velocity of a given mass, we can observe the evolution of the coefficients, where the trajectories appear to be quite regular (see Figure 10.4). Let's focus on the energy and its flow through the modes (Figure 10.5). The energy associated with a given mode n is the sum of

Figure 10.4 Coefficients $a^i(t)$ and $b^i(t)$ for $i=1$, 2, 3, and 4 respectively, for the case as in Figure 10.3: (a) time dependent, (b) phase diagram

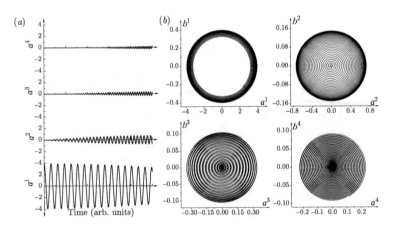

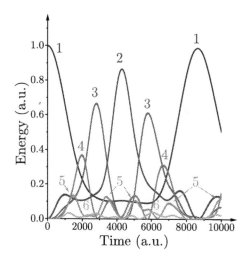

Figure 10.5 Solution for $N = 32$, $\alpha = 1$ and $\beta = 1/4$, originally obtained by Enrico Fermi, John Pasta, Stanisław Ulam and Mary Tsingou in 1955 (The numbers above the curves denote modes.)

potential and kinetic energies

$$E^n(t) = \sum_{i=1}^{N} U(y_i^n - y_{i-1}^n) + \sum_{i=1}^{N-1} \frac{1}{2} m v_i^{n2}, \qquad (10.1.16)$$

where the upper index is the mode number, and the lower index indicates the mass. We can also write the condition for energy conservation

$$\sum_{n=1}^{N-1} E^n(t) = \text{const} = \sum_{i=1}^{N} U(y_i^{10} - y_{i-1}^{10}). \qquad (10.1.17)$$

The computational procedure is now as follows. We solve the system of coupled differential equations (IVP – initial value problem), using one of the methods described in Project 3, while similarly to the coupled oscillators (Project 9), we recommend the Runge–Kutta method. For every fixed number of time steps (which speeds up the algorithm), we calculate the coefficients $a^n(t)$ and $b^n(t)$ from the dot products (10.1.13) and (10.1.15), which allow us to calculate the energies $E^n(t)$ with the use of (10.1.12), (10.1.14), and (10.1.16). That way we can obtain the time dependence of these quantities. The total energy as a function of time (which should be constant) will serve as a criterion for a proper choice of the time step (the optimal time step is the biggest time step that still keeps the conservation of energy).

Figure 10.6 Displacements
of masses at different
instances of time

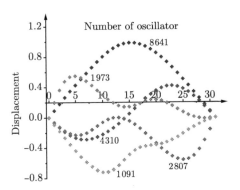

10.2 Exercises

1. Before proceeding to each of the exercises where the parameters of
the system are changed (the number of masses, coefficients of interactions), a convergence test with respect to the time step h should
be performed. The criterion for a proper choice is the conservation
of energy with respect to time. In order to practice the procedure,
try to find the largest h for which the energy is conserved. Use the
amplitude distribution for the first mode as initial positions (and zero
velocities).

2. For the system from the previous exercise perform a simulation of
motion. For a few chosen time instances plot the positions of the
mass (see Figure 10.6). Based on that plot, choose one relatively
complicated state of positions, and knowing the coefficients $a^n(t)$
plot in a single figure $y(t)$, and its component $\{a^n(t)\hat{y}^{n0}\}$.

3. For a chosen system plot the time dependence and phase portraits
of position and velocity of a chosen mass and $a^n(t)$ and $b^n(t)$ coefficients (see Figures 10.3 and 10.4). Repeat the exercise for three
cases: linear interaction only, non-linear interaction only, and both
interactions at the same time.

4. The most interesting aspect of the FPUT problem is the frequency
multiplication and energy transfer between modes. The first one was
present in Exercise 2 already, where the initial condition was a fundamental mode; higher-order modes appear after some time. Now
we investigate the time dependence of the energy components.
For a given system, plot the total energy $E(t)$ and energy components
$\{E^n(t)\}$ as a function of time (see Figure 10.5). Repeat the exercise
for three cases: linear interaction only, non-linear interaction only,
and both interactions at the same time.

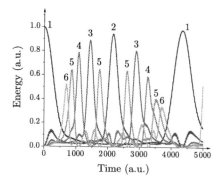

Figure 10.7 Results for $N = 32$, $a = 1$, $m = 1$, $\alpha = 1$, $\beta = 1$. Larger magnitude of the β parameter results in the emergence of higher-order modes

5. Investigate how the non-linear interaction parameter influences the speed of higher-order mode generation, their number and amplitude (see Figure 10.7). Suggest an appropriate numerical experiment. Add a third-order interaction and check its influence on a chosen example (without a detailed analysis).

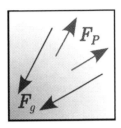

Project 11
Hydrogen Star

In Project 5 we discussed the problem of gravitation inside a star when the distribution of its mass density inside is known. In that case we applied the Numerov–Cowells algorithm to solve Poisson's equation in spherical coordinates – a boundary value problem. We can now go back to that problem and attempt to determine the mass density distribution.

11.1 Problem: Mass Density Distribution in a Cold Hydrogen Star

We already know how to determine the potential when a radial mass density distribution is known, so let us focus on the density. In order to do that we can use a condition for equivalence of forces acting on a mass dm at a distance r from the centre of the star and perpendicular to r. Two forces act on dm: a gravitational force and a force coming from the change in pressure as the distance is changed by dr (Figure 11.1). We can write this down as

$$a_G(r)\rho(r)drds = dP(r)ds, \qquad (11.1.1)$$

where $a_G(r)$ is a gravitational acceleration at a distance r from the centre, $\rho(r)$ is the mass density, and ds is an element of the surface perpendicular to the radius. The surface element ds can be eliminated from both sides of the equation, leaving the condition to be $a_G(r)\rho(r)dr = dP(r)$ (we are comparing the magnitudes of the forces here).

Figure 11.1 The condition for equivalence of forces acting on a mass dm

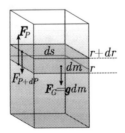

The quantities present in this equation need further analysis. The gravitational acceleration can be determined from the integral form of Gauss law

$$\oint_{S_r} a_G(r)\, ds = -4\pi\, GM_r, \tag{11.1.2}$$

where G is the gravitational constant, S_r is a spherical surface with a radius r, and M_r is the mass inside that sphere. To simplify further calculations we will assume $G = 1$.

Taking into account that the mass M_r is

$$M_r = \int_0^r \rho(r')4\pi r'^2\, dr', \tag{11.1.3}$$

and that a_G has the same value on the entire sphere S, from the integral form of Gauss law we have

$$4\pi r^2 a_G(r) = -4\pi \int_0^r \rho(r')4\pi r'^2\, dr', \tag{11.1.4}$$

$$a_G(r) = -\frac{1}{r^2} \int_0^r \rho(r')4\pi r'^2\, dr'. \tag{11.1.5}$$

On the right-hand side of the condition for force equivalence we will use the expression for $P(\rho)$ (from equation of state) for the gas that makes up the star, in order to express the differential dP through a partial derivative $\frac{\partial P}{\partial \rho}$ and a differential $d\rho$

$$dP = \frac{\partial P}{\partial \rho} d\rho. \tag{11.1.6}$$

Circling back to the condition for equivalence of forces we arrive at

$$-\frac{1}{r^2}\rho(r)dr \int_0^r \rho(r')4\pi r'^2\, dr' = \frac{\partial P}{\partial \rho} d\rho, \tag{11.1.7}$$

which leads to an integro-differential equation

$$\frac{d\rho}{dr} = -\frac{1}{r^2}\rho(r)\left(\frac{\partial P}{\partial \rho}\right)^{-1} \int_0^r \rho(r')4\pi r'^2\, dr'. \tag{11.1.8}$$

11.2 Numerical Method

In the discussed problem we are dealing with an integro-differential equation. It turns out, as we will show in this project, that the method used in Project 3 can also be used here. For solving numerically a differential equation

$$\frac{dy}{dx} = f(y,x), \qquad (11.2.1)$$

a uniform grid x_n with a constant h is introduced, which leads to a recursive scheme

$$y_{n+1} = y_n + h \int_{x_n}^{x_{n+1}} f(x,y)dx. \qquad (11.2.2)$$

We should note that the equation, even though it is an integro-differential equation, also has the form of 11.2.1. The function $f(x,y)$ in the integral needs to be extrapolated in the interval (x_n, x_{n+1}) based on the known previous points. We will use a linear extrapolation from points f_{n-1}, f_n. We can introduce a local variable $\xi = x - x_n$), then

$$f(\xi) = \frac{f_n - f_{n-1}}{h}\xi + f_n, \qquad (11.2.3)$$

and

$$y_{n+1} = y_n + \int_0^h \left(\frac{f_n - f_{n-1}}{h}\xi + f_n \right) d\xi. \qquad (11.2.4)$$

After the integration, we arrive at a second-order Adams–Bashforth scheme

$$y_{n+1} = y_n + \frac{1}{2}h(3f_n - f_{n-1}). \qquad (11.2.5)$$

In our equation, y is replaced with ρ, the independent variable is r, and the function f takes the form

$$f(r) = -\frac{1}{r^2}\rho(r) \left(\frac{\partial P}{\partial \rho} \right)_r^{-1} \int_0^r \rho(r')4\pi r'^2\, dr'. \qquad (11.2.6)$$

In the next step, the form of f_n after the discretisation $\{r_n = nh, n = 0, \ldots, N\}$ has to be found. In order to determine the integral in Eq. 11.2.6 we will use the trapezoid method

$$\int_0^r \rho(r')4\pi r'^2\, dr' = 4\pi \sum_{i=1}^n \frac{1}{2}h[\rho((i-1)h)(i-1)^2h^2 + \rho(ih)i^2h^2]. \qquad (11.2.7)$$

Therefore,

$$f_n = f(nh)$$

$$= -\frac{1}{n^2 h^2} \rho(nh) \left(\frac{\partial P}{\partial \rho}\right)_{hn}^{-1} 4\pi \sum_{i=1}^{n} \frac{1}{2} h[\rho((i-1)h)(i-1)^2 h^2 + \rho(ih)i^2 h^2],$$

$$(11.2.8)$$

and after simplification

$$f_n = -2\pi h \frac{\rho_n}{n^2} \left(\frac{\partial P}{\partial \rho}\right)_{hn}^{-1} \sum_{i=1}^{n} [\rho_{i-1}(i-1)^2 + \rho_i i^2]. \qquad (11.2.9)$$

Equation (11.2.9) can be used in a recursive scheme

$$\rho_{n+1} = \rho_n + \frac{1}{2} h(3f_n - f_{n-1}), \qquad (11.2.10)$$

beginning with $n = 1$. As we can see, it is necessary to use the initial values ρ_0 and ρ_1.

We should note that the density cannot change abruptly as we cross the centre of the star, which means that the derivative of the density with respect to r for $r = 0$ has to be equal zero, $d\rho/dr|_{r=0} = 0$. This condition is not fulfilled for the model density from Project 5, which is its significant flaw. Therefore, we can assume $\rho_0 = \rho_1$. We can also demonstrate that $f_0 = 0$ and subsequent values of f_n can be determined from 11.2.9.

We now need a non-recursive form of pressure $P(\rho)$. That pressure comes from the equation of state, and for an ideal gas it is linear. However, it turns out that the linear dependence is too weak to counteract a gravitational collapse. We note that as the density increases, the atoms come closer together, atomic orbitals start to overlap, and quantum interactions lead to a rapid increase in density. Here, we will propose an exponential form of $P(\rho)$.

$$P(\rho) = A(e^{B\rho^2} - 1), \qquad (11.2.11)$$

where A and B are parameters which influence the final result of $\rho(r)$ in one of the student exercises. In the limit of small density we obtain a linear dependence, like in an ideal gas; however, as the density increases, that increase is controlled by the constants A and B and can be sudden. A partial derivative that appears in Eq. 11.2.9 is of the form

$$\frac{\partial P(\rho)}{\partial \rho} = 2AB\rho e^{B\rho^2}. \qquad (11.2.12)$$

Therefore, we built a model to determine the mass density distribution in a star that has three parameters: the density in the centre of the star ρ_0 and two parameters A and B that describe the properties of the gas. The numerical experiments to be performed should focus on the influence of these parameters on the density distribution, a physical interpretation of that influence and the relation of the model to real life systems.

11.3 Exercises

1. Using the initial value for the density from the density model in Project 5, $\rho_0 = 1/(8\pi)$, choose the parameters A and B that describe the properties of the gas so that the density is close to the model one. Plot both densities in one figure and describe qualitative differences between them.

2. Try to make the model realistic by introducing the true value of the gravitational constant and by using SI units. Using the pressure as a function of volume per 1 atom, V_{cell}, obtained from density functional theory (DFT) (see Appendix A.9), find the relation between pressure and density $P(\rho)$, in order to determine the constants A and B from a fit to Eq. 11.2.11. Add also to the pressure versus density model (11.2.11) the term accounting for the equation of state of an ideal gas: $P = \frac{R}{\mu}T\rho$, where R is universal gas constant, μ is molar mass of hydrogen, and T is the temperature. Note that with this term also the derivative 11.2.12 has to be modified.

3. Plot the gravitational acceleration a_G (Eq. 11.1.5) as a function of the radius for different masses and temperatures, and determine the radius of the star R_S by finding the maximum value of the acceleration. Note that the mass of the star is determined by the density in the centre, ρ_0, and must be calculated from integral 11.1.3, at the upper limit determined by density approaching zero.

4. Using the realistic model of the star, determine the pressure at the centre of the star as a function of the star mass for a given temperature. The extrapolation of that dependence to zero gives the critical mass of the star M, that is the mass above which the system is stable (the gravitational force keeps the star), and below which the gas would disperse. Note that in these considerations also the escape speed of the hydrogen atom must be taken into account, that is the average velocity of the atom estimated from the formula $v_H = \sqrt{3k_BT/m}$ (where k_B is Boltzmann constant and m is atomic mass of hydrogen) should not exceed the escape speed $v_{esc} = \sqrt{2GM/R_S}$ (where G is gravitational constant and R_S can be taken as calculated previously radius of the star). This condition can

be used to establish the temperature range which can be considered for a given mass of the star.

5. Using literature data for hydrogen nuclear fusion reaction (pressure vs. temperature) discuss the possibility of such nuclear reaction in considered stars.

6. Assuming that the phase transition between gaseous and solid (metallic) hydrogen occurs at $400GPa$ (at $T = 0K$), find the radius of the metallic core of the star as a function of the mass of a hypothetical cold hydrogen star ($T = 0K$). For simplicity, assume that there is no change in density when the phase transition between hydrogen gas and metallic hydrogen occurs.

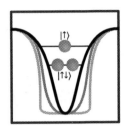

Project 12

Rectangular Quantum Well Filled with Electrons – The Idea of Self-Consistent Calculations

A quantum well, an example of which has already been discussed in Project 1, is an interesting problem relevant in real life applications. Semiconductor heterostructures, that is systems of layered semiconductor materials of varying properties (band gaps, effective masses, etc.), have a multitude of applications; therefore, it is not purely an academic exercise. In the direction perpendicular to that structure we observe a system of quantum wells with complicated electronic structures of discrete energy levels resulting from quantisation due to spatial confinement and electronic structure of the materials. In this project we partially address that problem by analysing a single quantum well (similar to that from Project 1) filled with electrons.

It is necessary to pay particular attention to the conditions required for such well to be formed. An empty quantum well, as in Project 1, is made of two infinite and infinitely thin dipole planes separated by a width a and of opposite dipole orientations. In that case, charge neutrality is present and, simultaneously, the plane introduces a sharp change in the potential to $-V_0$. If we were to fill that well with electrons, the situation would get more complicated. Charge neutrality has to be fulfilled but by introducing electrons (a certain amount of electrons per unit area), the well becomes negatively charged. One way to neutralise that charge is to introduce a uniform positive charge between the dipole planes, of equal charge per unit area. This model is called *jellium* and will be used in this project.

Let's start from an empty quantum well composed of infinite dipole planes filled with positively charged jellium. As we will soon realise, the shape will be different from that shown in Project 1. Utilising the algorithm of finding energy levels for a quantum well of any shape, discussed in Project 1, we will determine these levels and then populate them with electrons in numbers equivalent to the positive charge of the jellium. The presence of electrons will significantly modify the shape of the well compared to that for a single electron, since it now has to interact with a non-uniform electron gas. The problem is complicated on its own and is a basis for a whole field of studies, for example in density

functional theory (DFT); we, however, will greatly simplify the problem by considering only the Hartree interactions, that is the coulomb
interaction of a single electron with an average charge of the remaining
electrons. The calculations will be self-consistent. After determining
the energy levels in an empty well, we populate them with electrons
(respecting the Pauli's exclusion principle), therefore introducing a negative charge density to the positive charge density of the jellium. The
negative charge is evaluated from the formula

$$n_e(x) = -\sum_i \eta_i |\psi_i(x)|^2, \tag{12.0.1}$$

where η_i assumes values $0, 1, 2$, when the number of electrons in a
given state is $0, 1$, or 2, respectively (Pauli's exclusion principle). As
one can see, the summation covers occupied states only. Next, we
add the electron density $n_e(x)$ to the uniform density of jellium $n_j(x)$,
$n(x) = n_e(x) + n_j(x)$, and solve the Poisson's equation for the new
density

$$\frac{d^2 V(x)}{dx^2} = -4\pi n(x) \tag{12.0.2}$$

to find the new shape of the potential energy of the well $V(x)$. We then
determine the new energy levels, populate them with electrons, and
continue that loop until a certain property, such as the total energy, the
value of the ground state energy, or the electron density, converges.
An appropriate criterion for the convergence should be worked out
and implemented. Usually, when the differences between three to four
subsequent values of a chosen quantity (or residual of a function) do
not exceed an assumed small value, the loop stops (convergence is
achieved).

Here, we do not consider the electron self-interaction – occupying
the ground state with a single electron also modifies the potential, even
though the electron does not interact with itself.

12.1 Problem: Quantum Well Filled with Electrons and Charge Neutralising Jellium

Simple reasoning, even using just the integral form of the Gauss law,
leads to a conclusion that positive jellium introduced between two
infinite dipole planes is described with a parabolic potential inside
the system and a linear one outside. The electric field at the edge
of the jellium on both sides of the dipole layer has to be equal and
have the value of $E_o = 1/2N$ per a surface unit (in Hartree), where N is

Figure 12.1 Results of the self-consistent calculations. The dashed lines represent the results for the first iteration, and the solid lines represent the last iteration (after convergence is achieved), for a well of width 1, a potential 10 eV deep, and $N = 3$. The left panel shows the potential with the eigenenergies and their eigenfunctions. The right panel shows the total electron density for the same potential

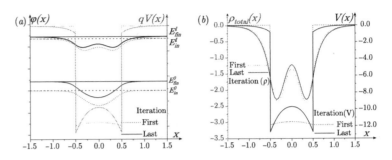

the expected number of electrons to be placed in the well. The direction of the field on both sides of jellium will be opposite.

The potential of an empty well can, therefore, be defined as

$$
V(x) = \begin{cases} -\dfrac{N}{2a}x^2 - V_o & \text{dla } -a/2 \le x \le a/2, \\ \dfrac{N}{a}x & \text{dla } x < -a/2, \\ -\dfrac{N}{a}x & \text{dla } x > a/2, \end{cases}
$$

where a is the width of the well (the distance between the dipole planes – the thickness of the jellium), N is the number of electrons that will occupy the states in the well, and V_o is the minimal depth of the well. The electric field at the well's edge is $-N/2, +N/2$, for negative and positive sides of the X axis. As a consequence of the self-consistent procedure, the potential outside of the well, initially linear, should flatten out, since the well is electrically neutral (Figure 12.1).

In the code, the algorithm described in Project 6 is used to determine the eigenstates of the quantum well. In the self-consistent procedure, the ground state energy is used as the convergence criterion.

12.2 Exercises

1. Analyse the *Hartree* code and run it for a chosen set of parameters.
2. Perform the calculation for different sets of parameters a, N, and V_o. Pay attention to how the potential changes with different values of N.
3. Perform the calculation for $N = 1$. Discuss the result.

Project 13

Time-Dependent Schrödinger Equation

Dawid Dworzański

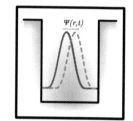

In this project, we discuss the problem of time evolution of a wave function in a given quantum well.

13.1 Problem: Time Evolution of a Wave Function in 2D Quantum Well

In previous projects we solved for stationary states for some quantum wells (1). These stationary states fulfil the equation

$$\hat{H}\psi_i(x,t) = \varepsilon_i \psi_i(x,t). \qquad (13.1.1)$$

In order to determine their evolution in time, we solve the time-dependent Schrödinger equation, which (in atomic units, $\hbar = m_e = e = 1$) reads

$$i\frac{d\psi(x,t)}{dt} = \hat{H}\psi(x,t). \qquad (13.1.2)$$

If the Hamiltonian is not explicitly dependent on time, the wave function can be written as a product of its position-dependent part and its time-dependent part

$$\psi_i(x,t) = \psi_i(x)\psi_i(t). \qquad (13.1.3)$$

Introducing the wave function to the Schrödinger equation 13.1.2 and utilising Eq. 13.1.1 we obtain

$$i\psi_i(x)\frac{d\psi(t)}{dt} = \varepsilon_i \psi_i(x)\psi_i(t), \qquad (13.1.4)$$

and

$$\frac{d\psi(t)}{dt} = -i\varepsilon_i \psi_i(t). \qquad (13.1.5)$$

The solution to that equation is an exponential function

$$\psi_i(t) = e^{-i\varepsilon_i t}. \qquad (13.1.6)$$

It is a complex number of magnitude 1 and phase $-\varepsilon_i t$, so the time evolution of stationary states is characterised with a uniform change of global phase at a speed proportional to the energy, since for a given x_0, only the phase of the wave function changes while the amplitude stays the same.

We can now determine the time evolution of any wave functions. In order to do that we will rely on the fact that the wave functions of stationary states form a complete basis in the Hilbert space; therefore, we can express any wave function as a linear combination of functions

$$\psi(x, t) = \alpha_1 \psi_1(x) + \alpha_2 \psi_2(x) + \alpha_3 \psi_3(x) + \cdots . \qquad (13.1.7)$$

To obtain the α_i coefficients we use the fact that the stationary function are orthonormal, that is

$$\int_{-\infty}^{\infty} \psi_i(x)^* \psi_j(x) \, dx = \delta_{i,j}, \qquad (13.1.8)$$

where $\delta_{i,j}$ is the Kronecker delta. We arrive at

$$\int_{-\infty}^{\infty} \psi_i(x) \psi(x) \, dx = \alpha_i . \qquad (13.1.9)$$

We can now determine the time evolution of any wave function (provided we know the stationary states of the hamiltonian) by decomposing the wave function into a linear combination of stationary functions 13.1.7 with the use of 13.1.9. Next, we complement each component with a time-dependent part due to 13.1.3 (notice that the above relations for stationary states are also fulfilled when we turn off the time-dependent part, and for $t = 0$ we get the stationary state basis)

$$\psi(x, t) = \alpha_1 \psi_1(x) e^{-i\varepsilon_1 t} + \alpha_2 \psi_2(x) e^{-i\varepsilon_2 t} + \alpha_3 \psi_3(x) e^{-i\varepsilon_3 t} + \cdots .$$
$$(13.1.10)$$

In the numerical calculations we have to limit the number of stationary functions to those of the lowest energies. In order to determine α_i we will use the numerical integration methods described in Project 2.

This method is simple and demonstrates the idea behind the reason for time evolution of wave functions. However, the task of determining the stationary states for a given potential, necessary to apply the method, is not simple. For that reason, we propose a different method which relies on direct discretisation of time-dependent Schrödinger equation (13.1.2), with the use of the three-point formulas for derivatives. The wave function is discretised in space with a step of h_x and in time with a step of h_t. Then, for $x, t \in \mathbb{Z}$ we can write

$$\psi_{x,t} = \psi(x h_x, t h_t) . \qquad (13.1.11)$$

We separate the Hamiltonian into kinetic and potential parts in the Schrödinger equation

$$i\frac{d\psi}{dt} = \hat{H}\psi = \frac{\hat{p}^2}{2}\psi + \hat{V}\psi = -\frac{1}{2}\frac{d^2\psi}{dx^2} + V(x)\psi, \qquad (13.1.12)$$

and using the formulas for numerical derivative from Sec. 2.3 we get

$$\frac{\psi_{x,t+1} - \psi_{x,t-1}}{2h_t} = \frac{i}{2}\frac{\psi_{x-1,t} + \psi_{x+1,t} - 2\psi_{x,t}}{h_r^2} - iV(x)\psi_{x,t}. \quad (13.1.13)$$

We can solve it for $\phi_{x,t+1}$, which leads us to

$$\psi_{x,t+1} = \psi_{x,t-1} + i\frac{h_t}{h_x^2}(\psi_{x-1,t} + \psi_{x+1,t} - 2\psi_{x,t}) - iV(x)\psi_{x,t}2h_t.$$

$$(13.1.14)$$

It is important to notice that $\frac{h_t}{h_x^2}$ is present in that equation; therefore, in order to obtain sufficiently low order, h_t has to be at least three orders lower than h_x.

In addition, a problem with boundary condition appears. We can assume that the values on the boundaries are equal to zero and do not change. This corresponds to a situation in which an infinite barrier is present at the boundary. Alternatively, we can use periodic boundary conditions if we assume that the next value of the wave function to the left of the left boundary is the first value of the wave function at the right boundary, and vice versa. By iterating the obtained formula we obtain the wave function at any given time. In order to ensure the simulation is free of numerical errors and physically correct, we can investigate the expected value of the energy

$$\langle \hat{H} \rangle = \int_{-\infty}^{\infty} \psi^*(x)\,\hat{H}\,\psi(x)\,dx = \int_{-\infty}^{\infty} \psi^*(x)\left(-\frac{d^2\psi}{dx^2} + V(x)\psi(x)\right)dx,$$

which should be conserved.

Using this method we can demonstrate many phenomena in quantum physics.

For sake of clarity, the plots show only 10% of the potential. We also use the concept of a wave packet, that is a wave function in the form

$$\psi(x) = C\,e^{ikx}\,e^{-(x-x_0)^2/2\sigma^2}, \qquad (13.1.15)$$

where C is a normalisation constant which can be obtained for any function as an inverse of the integral of the function, in this case $C = \frac{1}{\sigma\sqrt{2\pi}}$, k is the expectation value of the momentum, x_0 is the expectation value of the position, and σ is the standard deviation of the position.

13.2 Exercises

Mandatory

1. Run the WAVE_BASE code, which decomposes a given wave function into stationary states of an infinite rectangular quantum well (Figure 13.1(a)), then run the WAVE code, which uses finite differences, and compare the evolution of the wave functions for both approaches for different initial wave functions, in particular stationary states of an infinite square well, wave packet of momentum equal to zero (Figure 13.1(b)), and a wave packet of momentum not equal to zero (Figure 13.1(c)).

2. Run the WAVE code for a parabolic potential for a wave packet at equilibrium (Figure 13.2(a)) and a non-zero momentum (Figure 13.2(b)). Plot the expectation value of the position

$$\langle \hat{x} \rangle = \int_{-\infty}^{\infty} \psi^*(x)\, x\, \psi(x)\, dx = \int_{-\infty}^{\infty} x\, |\psi(x)|^2\, dx$$

as a function of time.

3. For a quantum harmonic oscillator, plot the expectation value of energy as a function of time. Check how the kinetic and potential parts change, that is $\int_{-\infty}^{\infty} -\psi^*(x)\frac{d^2\psi}{dx^2}\, dx$ and $\int_{-\infty}^{\infty} \psi^*(x)V(x)\psi(x)\, dx$.

4. Investigate the phenomenon of quantum tunnelling (Figure 13.3). Check how much of the wave packet reflects off of the barrier, and

Figure 13.1 For a zero potential (placed in a wide infinite quantum well) the above scenarios can be considered (a) infinite quantum well results in a stationary probability distribution, (b) a zero momentum results in the wavefunction spreading, that is increasing the uncertainty of position, and (c) a non-zero momentum results in the wavefunction spreading an moving

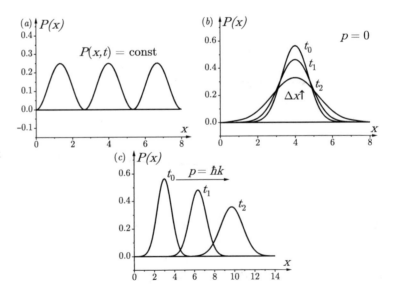

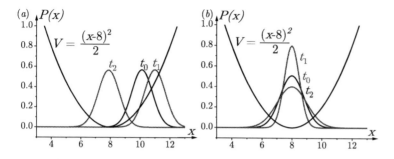

Figure 13.2 For a parabolic potential we obtain a quantum harmonic oscillator: (a) choosing a non-zero momentum and $\sigma^2 = 0$, the obtained state oscillates around the equilibrium position without changing its shape (sloshing modes), (b) setting $\sigma^2 = 2$ and $p = 0$, we observe an alternating spreading and tightening of the state, without changing its position (breathing modes)

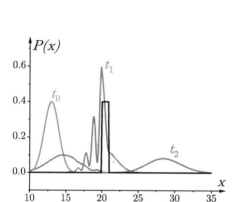

Figure 13.3 For a potential barrier we can observe quantum tunnelling. A wave packet after a collision with the barrier partially reflects and partially tunnels through it

how much tunnels through it, as a function of width and thickness of the barriers and the wave packet's momentum.

Advanced

1. Modify the WAVE_BASE code so that using stationary functions for a parabolic potential it calculates the time evolution of any wave function for this potential.

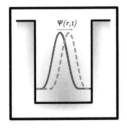

Project 14
Poisson's Equation in 2D

In the projects presented until now, the systems had high symmetry (cylindrical or spherical). Owing to this, even very complicated equations, such as the Laplace's equation, could be greatly simplified. For example, in the problem described in Project 5, the symmetry was spherical, while in Project 8 we took advantage of cylindrical symmetry to search for a quasi-1D potential. In this project we will search for a potential for a given distribution of charges in a two-dimensional system which does not have any symmetry. This is an extension of the project from Project 8. We also use the method of minimising a functional, in particular the Gauss–Seidel method of iterative minimisation. The Poisson's equation will be adjusted to a two-dimensional case, which means neglecting one partial derivative in Cartesian coordinates. The resulting code will be capable of modelling typical electrostatic systems such as an electric dipole, systems of charges, parallel plates of a capacitor, or a Faraday's cage. One should remember, however, that a point charge in this representation is in fact an infinite line of certain charge density, and a segment represents an infinite plate of width equal to the length of the segment. It is interesting to note that the interaction between such point charges has no $1/r^2$ character anymore but $1/r$. This should be kept in mind while analysing the results of the simulations.

14.1 Problem: Variational Computational Approach to a 2D Electrostatic System

In this project we will analyse a two-dimensional system, for which Poisson's equation (8.1.1) takes the form

$$\left(\frac{\partial^2}{\partial x^2} + \frac{\partial^2}{\partial y^2} \right) \phi(x, y) = -4\pi \rho(x, y), \qquad (14.1.1)$$

which, together with the boundary conditions, leads to a boundary value problem (analogous to the one discussed in Project 5). The functional (8.1.2), written in two dimensions is

$$F[\phi] = \int_S d^2\mathbf{r} \left[\frac{1}{2} \left[\left(\frac{\partial \phi(x,y)}{\partial x} \right)^2 + \left(\frac{\partial \phi(x,y)}{\partial y} \right)^2 \right] - 4\pi \rho(x,y)\phi(x,y) \right]. \tag{14.1.2}$$

The fundamental question of electrostatics can be summarised as *What is the potential of the electric field for a given distribution of charges?* It is worth spending some time on looking at some peculiar features of solutions to this problem. An important, non-trivial issue, known to those well versed in this field of physics, is the uniqueness of the results. This means that we would like to know whether Poisson's equation is uniquely (unequivocally) given by the boundary conditions. If that was not the case, we could have a situation in which the functional would have not one, but multiple global minima. This question has to be answered before we proceed with the discussion of the numerical approach to solving the problem, as this is of a deeper character than being just a numerical problem to solve, and is rooted in an analytical approach to searching for the solution for the field. We should expect that the uniqueness of the solution does not result from the method chosen to search for it, but from the fundamental laws that define it.

It should be clear that the determined potential is always unique. Otherwise we could suspect that the world, as any other system, exists in one of many metastable states in terms of the distribution of the electrostatic field, and due to some hard to determine factors could switch between them, which obviously does not happen. The uniqueness theorems prove that, in the form of two laws. Therefore, we expect that any initial conditions should always lead to the same solution, which the reader should verify as an exercise. Both these theorems are presented in depth in the Appendix (A.6).

14.2 Numerical Method: Finite Elements (FE) Method

Similarly to previously discussed problems, we start by discretisation of the space we work in, which means we introduce a grid of points on which the calculations are performed. We use a rectangular grid, in which the grid parameters in the two directions may differ (h_x, h_y). As a result of discretisation the functional (14.1.2) takes the form

$$F[\phi] = \frac{h_x h_y}{2} \sum_{i=2}^{N_i-1} \sum_{j=2}^{N_j-1} \left[\frac{(\phi_{ij} - \phi_{i-1j})^2}{h_x^2} + \frac{(\phi_{ij} - \phi_{ij-1})^2}{h_y^2} \right] -$$

$$-h_x h_y \sum_{i=2}^{N_i-1} \sum_{j=2}^{N_j-1} 4\pi \rho_{ij} \phi_{ij}.$$

(14.2.1)

After discretisation, the functional becomes a multi-variable function, $F(\phi) \rightarrow F(\phi_{ij})$, and the condition of zeroing of δF reduces to a condition of zeroing of partial derivatives, $\partial F/\partial \phi_{ij} = 0$. Using that condition, we arrive at

$$\phi_{ij} = \frac{h_x^2 h_y^2 4\pi \rho_{ij} + h_y^2(\phi_{i-1j} + \phi_{i+1j}) + h_x^2(\phi_{ij-1} + \phi_{ij+1})}{2(h_x^2 + h_y^2)}, \quad (14.2.2)$$

which in a special case of a square grid ($h_x = h_y \equiv h$) simplifies to

$$\phi_{ij} = \frac{1}{4} \left(h^2 4\pi \rho_{ij} + \phi_{i-1j} + \phi_{i+1j} + \phi_{ij-1} + \phi_{ij+1} \right). \quad (14.2.3)$$

Detailed derivation of the above relations is presented in the Appendix (A.7. Interestingly, the above relations could have also been obtained in a slightly simpler way, by discretising the Poisson's equation (5.1.2), or directly from an expression for an average potential (A.6.1) in two dimensions, which for $\rho(\mathbf{r}) = 0$ is just a regular arithmetic mean, $\phi_{ij} = \frac{1}{4}(\phi_{i-1j} + \phi_{i+1j} + \phi_{ij-1} + \phi_{ij+1})$. It is important to note that the boundary condition can be imposed not only on the boundaries but also inside the considered area through fixed values in chosen nodes of the grid. This way different electrostatic systems can be simulated (capacitors, Faraday cage, point charges).

To sum up, the algorithm does not fundamentally differ from that used for the cylindrical capacitor (a quasi one-dimensional system):

i. Start with some initial function ϕ_{ij} (e.g. a surface $\phi(x_i, y_j) = a$, between $\phi(0, y_j), \phi(x_i, 0), \phi(x_{N_i}, y_j), \phi(x_i, y_{N_j})$) (excluding certain points of previously set values).

ii. Calculate the new ϕ_{ij} using Eq. 14.2.2 at each point on the grid x_i, y_j, except the boundaries.

iii. For the new $\{\phi_{ij}\}$, calculate the value of the functional $F[\phi]$ (14.2.1).

iv. Repeat the procedure until the value stabilises by reaching a $F[\phi^{new}] - F[\phi^{old}]$ less than ε.

The procedure can be slightly modified by mixing the potentials ϕ_{ij}

$$\phi'_{ij} = \phi^{new}_{ij}\omega + \phi^{old}_{ij}(1 - \omega), \qquad\qquad (14.2.4)$$

where ϕ^{new}_{ij} is the new solution calculated with the use of the old value (14.2.2) ϕ^{old}_{ij}.

However, the two-dimensional case introduces a new freedom in the choice of the sequence of actions in which the values are updated. We take the new values of ϕ from neighbouring nodes, and it may then turn out that the sequence of actions in which the values are updated (randomly, in rows, in columns, both at the same time), can influence the efficiency of calculations. As an exercise, the reader should check different possibilities.

14.3 Exercises

Testing of the Algorithm

1. Analyse the 2POISSON code and test different boundary conditions and control parameters.
2. Investigate the effect of mixing the potentials. Find the range of the ω parameter for which the algorithm converges and check when the convergence is achieved the fastest. Note that the convergence may be dependent on the distribution of charges.
3. Perform fully converged calculations for boundary conditions for which the analytical solution is known, and compare the results to this analytical solution.
4. Verify the uniqueness theorem. In order to do that determine the potential for a given charge distribution with given fixed boundary conditions for different initial conditions of the iterative procedure.

Application of the Model
5. Consider two, metal, infinite (in the z direction), grounded (potential equals zero) planes $y = 0$ and $y = +a$ of a given width l, connected by a third plane $x = 0$, on which the potential $V(y)$ is given (Figure 14.1) Find the potential between the planes for
   (a) $V(y) = V_0$
   (b) $V(y) = V_0 y$
   (c) $V(y) = -V_0 y(y - a)$
6. Consider two, metal, infinite (in the z direction) planes of width l at $y = 0$ i $y = +a$ with a constant potential $+V$, which are

Figure 14.1 The model considered in this exercise – two connected planes

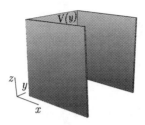

Figure 14.2 The model considered this exercise – two connected planes by two strips

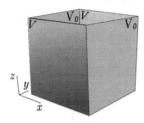

Figure 14.3 The box proposed to study in this exercise

    (a) isolated

    (b) connected by two planes with a constant potential V_0 (Figure 14.2):

        i. $0 < V_0 < V$

        ii. $V < V_0$

        iii. $0 < V_0 < V$ on the first one $V < V_0$ on the second one

7. Metal, grounded, rectangular cuboid box, covered by a lid insulated from the walls and uniformly charged with a potential V (Figure 14.3). Find the potential inside the box, around its centre (in the z direction).

8. Consider a model of a two-dimensional crystal of 10×10 positively charged ions of a metal ($+Q$), and alternating positive and negative ions (notice that it is actually a system of infinite linear charges). Investigate into the distribution of the potential in such a system.

9. Find the distribution of charges, which would result in a point of persistent equilibrium for a freely moving charged mass somewhere

between the points in which the charges reside. Is such a situation possible at all?

10. Design and simulate simple electrostatic systems such as an electric dipole, quadrupole, flat capacitor, Faraday's cage. (As mention earlier the code simulates quasi two-dimensional systems, therefore a single charge is represented here as an infinite linear charge.) As a boundary condition for the whole area of the simulation, a constant value has to be assumed, for example zero.

Appendix A
Supplementary Materials

A.1 Euler Representation of a Complex Number

A complex number z is an expression $z = a + ib$, where a, b are real numbers: $a = Re(z)$ is the real part, $b = Im(z)$ the imaginary part, and $i = \sqrt{(-1)}$ is a unit of imaginary part. A complex number conjugate to z is $z^* = a - ib$ and modulus of a complex number $|z| = \sqrt{(zz^*)} = \sqrt{(a^2 + b^2)}$.

Using the above quantities we can represent a complex number in the form

$$z = |z|(\frac{a}{\sqrt{(a^2 + b^2)}} + i\frac{b}{\sqrt{(a^2 + b^2)}}). \qquad (A.1.1)$$

The fractions appearing in the above expression can be represented as cos and sin functions of a right triangle, thus we can write

$$z = |z|(\cos(\theta) + i\sin(\theta)). \qquad (A.1.2)$$

This form suggests a useful graphical representation of a complex number, the so-called Argand diagram (Figure A.1). This is just a rectangular coordinate system on a complex plane, in which we represent the real part on a horizontal axis, and imaginary part of the vertical axis. The point (vector) on a plane defined in this way represents a complex number, and its representation via the angle θ (between the horizontal axis and the position vector) and the length of the vector ($|z|$) are formally analogous to polar coordinates associated with the Cartesian coordinates on a plane.

Moreover, an interesting mathematical fact can be noted, which leads to Euler (exponential) representation of a complex number $z = |z|e^{i\theta}$. Namely, it turns out that if we expand this exponential function into a power series then the series splits into a real part (for even powers of i) and an imaginary part (for odd powers of i). These two series appear to be power expansions of cos and sin functions of θ:

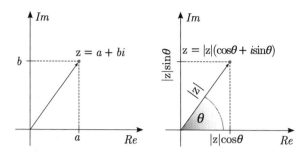

$$e^{i\theta} = 1 + i\theta + \frac{(i\theta)^2}{2} + \frac{(i\theta)^3}{3!} + \frac{(i\theta)^4}{4!} + \frac{(i\theta)^5}{5!} + \frac{(i\theta)^6}{6!} + \frac{(i\theta)^7}{7!} + \cdots$$

$$= 1 + i\theta - \frac{\theta^2}{2} - i\frac{\theta^3}{3!} + \frac{\theta^4}{4!} + i\frac{\theta^5}{5!} - \frac{\theta^6}{6!} - i\frac{\theta^7}{7!} + \cdots$$

$$= \left(1 - \frac{\theta^2}{2} + \frac{\theta^4}{4!} - \frac{\theta^6}{6!} + \cdots\right) + i\left(\theta - \frac{\theta^3}{3!} + \frac{\theta^5}{5!} - \frac{\theta^7}{7!} + \cdots\right)$$

$$= \cos\theta + i\sin\theta.$$

Looking at Eq. A.1.2 we find out that a complex number can be represented in the Euler representation

$$z = |z|e^{i\theta}. \tag{A.1.3}$$

Such a form is very convenient in the physics of oscillations and waves, when used with the superposition principle. Additionally, there are methods for solving linear differential equations that utilise the Euler representation of a complex function as a trial solution.

Here, to show advantages of Euler representation we will derive well-known relations for cos and sin functions of a sum of angles.

On the one hand, we have

$$e^{i(\alpha+\beta)} = \cos(\alpha+\beta) + i\sin(\alpha+\beta),$$

on the other hand,

$$e^{i(\alpha+\beta)} = e^{i\alpha}e^{i\beta} = (\cos(\alpha) + i\sin(\alpha))(\cos(\beta) + i\sin(\beta)),$$

$$= (\cos(\alpha)\cos(\beta) - \sin(\alpha)\sin(\beta)) + i(\cos(\alpha)sin(\beta)$$

$$+ \sin(\alpha)\cos(\beta)).$$

By comparing the real and imaginary parts we get the following relations

$$\cos(\alpha + \beta) = \cos(\alpha)\cos(\beta) - \sin(\alpha)\sin(\beta),$$
$$\sin(\alpha + \beta) = \cos(\alpha)\sin(\beta) + \sin(\alpha)\cos(\beta).$$

A.2 Local Representation of a Function as a Power Series

A.2.1 Taylor Series and Polynomials

An infinitely differentiable function $f(x)$ can be described as a following series (called the *Taylor series*)

$$f(x) = \sum_{k=0}^{\infty} \frac{f^{(k)}(x_0)}{k!}(x - x_0)^k$$

$$= f(x_0) + f'(x_0)(x - x_0) + \frac{f''(x_0)}{2!}(x - x_0)^2 + \dots \quad \text{(A.2.1)}$$

Taylor polynomials of n-th degree can be constructed by truncating the Taylor series (Eq. A.2.1) at the n-th term (i.e. by replacing the ∞ in the summation by n):

$$f(x) = P_n(x) + R_n(x), \quad \text{(A.2.2)}$$

where $P_n(x)$ is the n-th Taylor polynomial

$$P_n(x) = \sum_{k=0}^{n} \frac{f^{(k)}(x_0)}{k!}(x - x_0)^k, \quad \text{(A.2.3)}$$

and $R_n(x)$ is the remainder or the truncation error, which may be expressed as

$$R_n(x) = \frac{f^{(n+1)}(\xi(x))}{(n+1)!}(x - x_0)^{n+1}. \quad \text{(A.2.4)}$$

The remainder term may be used to estimate the uncertainty of a method derived from Taylor polynomials.

A.2.2 Taylor Series in Two Variables

Although not used in this book, it is worth to know how the Taylor series in two variables looks like, which can be easily extended to three and

more variables. It is based on the same notion: however, due to the introduction of a second variable the terms become much more complex, and
the general formula is

$$
f(x,y) = \sum_{k=0}^{\infty} \left[\frac{1}{k!} \sum_{j=0}^{k} \binom{k}{j} \frac{\partial^{(k)} f(x_0, y_0)}{\partial x^{(k-j)} \partial y^{(j)}} (x - x_0)^{k-j} (y - y_0)^j \right]
$$

$$
= f(x_0, y_0) + \left[(x - x_0) \frac{\partial f(x_0, y_0)}{\partial x} + (y - y_0) \frac{\partial f(x_0, y_0)}{\partial y} \right]
$$

$$
+ \left[\frac{(x - x_0)^2}{2!} \frac{\partial^2 f(x_0, y_0)}{\partial x^2} + (x - x_0)(y - y_0) \frac{\partial^2 f(x_0, y_0)}{\partial x \partial y} \right.
$$

$$
+ \left. \frac{(y - y_0)^2}{2!} \frac{\partial^2 f(x_0, y_0)}{\partial y^2} \right] + \left[(y - y_0) \frac{\partial f(x_0, y_0)}{\partial y} \right] + \ldots,
$$

$$
(A.2.5)
$$

where $\binom{k}{j}$ is the binomial coefficient defined as

$$
\binom{k}{j} = \frac{k!}{(k-j)! \, k!}. \tag{A.2.6}
$$

A.2.3 Lagrange Polynomials

An interesting alternative to Taylor expansion, of constructing polynomials passing through a set of points, is the Lagrange polynomial. The
general formula may be written as

$$
L(x) = \sum_{j=0}^{n} \left(y_j \prod_{\substack{k=1 \\ k \neq j}}^{n} \frac{x - x_k}{x_j - x_k} \right)
$$

$$
= \frac{(x - x_1)(x - x_2) \ldots (x - x_n)}{(x_0 - x_1)(x_0 - x_2) \ldots (x_0 - x_n)} y_0 \tag{A.2.7}
$$

$$
+ \frac{(x - x_0)(x - x_2) \ldots (x - x_n)}{(x_1 - x_0)(x_1 - x_2) \ldots (x_1 - x_n)} y_1
$$

$$
+ \ldots + \frac{(x - x_1)(x - x_2) \ldots (x - x_n)}{(x_0 - x_1)(x_0 - x_2) \ldots (x_0 - x_n)} y_n.
$$

The derivation of Lagrange polynomials is fairly simple.

It is worth to mention that the Lagrange polynomial for a given set
of points is unique and its uniqueness, although not explicitly shown in
this text, can be proven and found in a number of textbooks.

A.3 Wilberforce's Pendulum

Lagrangian of Wilberforce's pendulum has the form

$$L = \frac{1}{2}m\dot{z}^2 + \frac{1}{2}I\dot{\theta}^2 - \frac{1}{2}kz^2 - \frac{1}{2}\delta\theta^2 - \frac{1}{2}\epsilon z\theta. \qquad (A.3.1)$$

We use Euler–Lagrange equation

$$\frac{d}{dt}\left(\frac{\partial L}{\partial \dot{x}}\right) - \frac{\partial L}{\partial x} = 0, \qquad (A.3.2)$$

to obtain equations of motion

$$\frac{d}{dt}(m\dot{z}) + kz + \frac{1}{2}\epsilon\theta = 0, \qquad (A.3.3)$$

$$\frac{d}{dt}\left(I\dot{\theta}\right) + \delta\theta + \frac{1}{2}\epsilon z = 0, \qquad (A.3.4)$$

$$m\ddot{z} + kz + \frac{1}{2}\epsilon\theta = 0, \qquad (A.3.5)$$

$$I\ddot{\theta} + k\theta + \frac{1}{2}\epsilon z. \qquad (A.3.6)$$

A.4 Dispersion Relation for FPU Problem

To receive dispersion relation for an infinite chain of interacting masses (Project 10) we start with Newton's equation for i-th mass

$$m\frac{\partial^2 y_i}{\partial t^2} = \alpha(y_{i+1} + y_{i-1} - 2y_i), \qquad (A.4.1)$$

in which we introduce a solution in a waveform

$$y_i = Ae^{i(\omega t - kx_i)}, \qquad (A.4.2)$$

This substitution leads to

$$- m\omega^2 y_i = \alpha(y_{i+1} + y_{i-1} - 2y_i). \qquad (A.4.3)$$

We want to determine ω, so we note that

$$\frac{y_{i+1}}{y_i} = \frac{Ae^{i(\omega t - kx_{i+1})}}{Ae^{i(\omega t - kx_i)}} = e^{-ik(x_{i+1} - x_i)} = e^{-ika}, \qquad (A.4.4)$$

where we use lattice constant a as a distance between two masses in equilibrium points.

We use this result when dividing the previous equation by y_i

$$- m\omega^2 = \alpha(e^{-ika} + e^{ika} - 2) = \alpha(2\cos(ka) - 2) = 2\alpha[\cos(ka) - 1], \qquad (A.4.5)$$

which leads to the dispersion relation

$$\omega(k) = \sqrt{\frac{2\alpha}{m}[1 - \cos(ka)]}. \tag{A.4.6}$$

A.5 Equivalence of Variational Principle and Poisson's Equation in Electrostatics

To prove the equivalence of the Poisson's differential equation and the variational principle, let us consider the functional

$$F[\phi] = \int_V d^3\mathbf{r} \left[\frac{1}{2}(\nabla\phi(\mathbf{r}))^2 - 4\pi\rho(\mathbf{r})\phi(\mathbf{r})\right]. \tag{A.5.1}$$

and its change with respect to potential variation $\delta\phi(\mathbf{r})$, which, according to the variational principle, should be zero

$$\delta F = \int_V d^3\mathbf{r}\left[\nabla\phi(\mathbf{r}) \cdot \nabla\delta\phi(\mathbf{r}) - 4\pi\rho(\mathbf{r})\delta\phi(\mathbf{r})\right]. \tag{A.5.2}$$

Using the identity

$$\nabla \cdot (ab) = (\nabla a) \cdot \mathbf{b} + a(\nabla \cdot \mathbf{b}), \tag{A.5.3}$$

in which we substitute $a = \delta\phi(\mathbf{r})$ and $\mathbf{b} = \nabla\delta\phi(\mathbf{r})$, we obtain another identity

$$\nabla \cdot (\delta\phi(\mathbf{r})\nabla\phi(\mathbf{r})) = \nabla\delta\phi(\mathbf{r}) \cdot \nabla\phi(\mathbf{r}) + \delta\phi(\mathbf{r})\nabla \cdot (\nabla\phi(\mathbf{r}))$$
$$= \nabla\delta\phi(\mathbf{r}) \cdot \nabla\phi(\mathbf{r}) + \delta\phi(\mathbf{r})\nabla^2\phi(\mathbf{r}),$$

which when substituted to the functional change leads to

$$\delta F = \int_V d^3\mathbf{r}\left[\nabla \cdot (\delta\phi(\mathbf{r})\nabla\phi(\mathbf{r})) - \delta\phi(\mathbf{r})\nabla^2\phi(\mathbf{r}) - 4\pi\rho(\mathbf{r})\delta\phi(\mathbf{r})\right]$$
$$= \int_V d^3\mathbf{r}\left[\nabla \cdot (\delta\phi(\mathbf{r})\nabla\phi(\mathbf{r}))\right] - \int_V d^3\mathbf{r}\left[\delta\phi(\mathbf{r})\left(\nabla^2\phi(\mathbf{r}) + 4\pi\rho(\mathbf{r})\right)\right].$$

The first integral in the result can be evaluated using Gauss's theorem

$$\int_V d^3\mathbf{r}(\nabla \cdot \mathbf{F}) = \oint_S d^2\mathbf{r}(\mathbf{F} \cdot \mathbf{n}), \tag{A.5.4}$$

where the left side is the volume integral and the right one is the surface integral over volume V boundary. Thus

$$\int_V d^3\mathbf{r}\left[\nabla \cdot (\delta\phi(\mathbf{r})\nabla\phi(\mathbf{r}))\right] = \oint_S d^2\mathbf{r}\left[\delta\phi(\mathbf{r})\nabla\phi(\mathbf{r})\right] \cdot \mathbf{n}.$$

Owing to the fact that over the surface S $\delta\phi(\mathbf{r})$ is zero, the integral is equal to zero too.

Finally, we get

$$\delta F = -\int_V d^3\mathbf{r} \left[\delta\phi(\mathbf{r}) \left(\nabla^2\phi(\mathbf{r}) + 4\pi\rho(\mathbf{r}) \right) \right].$$

The above equation proves the equivalence of the Poisson's differential equation and the variational principle. Namely, if the differential equation is fulfiled, then the functional is stationary $\delta F = 0$, and if the functional is stationary $\delta F = 0$, then the Poisson's equation must be fulfilled because $\delta\phi(\mathbf{r})$ is non-zero everywhere except at the boundary.

A.6 First and Second Uniqueness Theory

The following theories allow to prove that solutions of Laplace's (first *u.t.*) and Poisson's (second *u.t.*) equations for given boundary conditions are uniquely determined.

A.6.1 First Uniqueness Theory

Let us consider a space in which certain boundary conditions for potential are given. From electrodynamics it is known that at every point the value of $\phi(\mathbf{r})$ is equal to average value of $\phi(\mathbf{r})$ over a spherical surface of radius R around this point (Figure A.2)

$$\phi(\mathbf{r}) = \frac{1}{4\pi R^2} \int_S d^2\mathbf{r}[\phi(\mathbf{r})]. \tag{A.6.1}$$

Notice that every possible potential cannot have extreme (maximum or minimum) inside considered space beyond the boundaries. Also, average value cannot be higher or lower than any of the values entering the averaging formula. Specially, if the boundary values are equal to zero, then the only possible solution is $\phi(\mathbf{r})$, equal to zero in whole space.

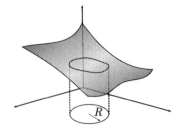

Figure A.2 Graphic representation equation (A.6.1) in two dimensions

Now consider any boundary conditions and suppose that there exist two solutions $\phi_1(\mathbf{r})$ and $\phi_2(\mathbf{r})$. Both must fulfill Laplace's equation

$$\nabla^2 \phi_1(\mathbf{r}) = 0,$$
$$\nabla^2 \phi_2(\mathbf{r}) = 0.$$

The potential

$$\phi_3(\mathbf{r}) \equiv \phi_1(\mathbf{r}) - \phi_2(\mathbf{r}), \qquad (A.6.2)$$

must fulfil Laplace's equation too

$$\nabla^2 \phi_3(\mathbf{r}) = \nabla^2(\phi_1(\mathbf{r}) - \phi_2(\mathbf{r})) = \nabla^2 \phi_1(\mathbf{r}) - \nabla^2 \phi_2(\mathbf{r}) = 0.$$

For points at boundary $\phi_1(\mathbf{r}) = \phi_2(\mathbf{r})$, so $\phi_3(\mathbf{r}) = 0$, but as we have noticed previously, if the potential boundary value is zero, then the only possible result is $\phi_3(\mathbf{r})$, equal to zero for every point, which implies $\phi_1(\mathbf{r}) = \phi_2(\mathbf{r})$ in the whole space.

A.6.2 Second Uniqueness Theory

Until now we were considering a space without charges inside; they were present only at the boundary and were accounted for via appropriate boundary conditions. Now, we additionally assume that there exists an arbitrary charge distribution $\rho(\mathbf{r})$ inside. Thus, we consider a finite electrostatic system, but this time we will start from Poisson's law for the electric field strength $\mathbf{E}(\mathbf{r})$ – (the Gauss law, Figure A.3) in the forms: differential

$$\nabla \cdot \mathbf{E}(\mathbf{r}) = 4\pi \rho(\mathbf{r}), \qquad (A.6.3)$$

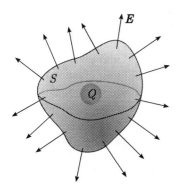

Figure A.3 Graphic representation of Gauss's law

and integral

$$\oint_S \mathbf{ds} \cdot \mathbf{E}(\mathbf{r}) = 4\pi Q, \tag{A.6.4}$$

where in the integrand there is a dot product of an oriented surface element and the electric field strength. The integral states that a total electric field flux through a closed surface is equal to the total charge enclosed in the surface. In its differential form we say that the charge is a source of the electric field.

The solution of the Poisson's equation (A.6.3) can be divided into two components: a solution to the uniform equation, that is the Laplace one, together with the boundary conditions, and a particular solution dependent on the charge distribution, for which the only boundary condition is the disappearing electric field at infinity. The general solution is a superposition of those two components $E = E^o + E^s$.

Consider now the issue of uniqueness of the general solution. In the previous section, we proved that the contribution from Laplace's equation (potential, thus also the field) is uniquely specified, it suffices then to prove that the particular solution E^s is also unique. For this purpose we will use the integral form of the Gauss law.

Suppose there exist two different solutions E_1^s and E_2^s for the same charge density distribution $\rho(r)$. For an arbitrary closed surface S within the considered space we have

$$\oint_S \mathbf{ds} \cdot \mathbf{E}_1^s(S) = 4\pi Q,$$

$$\oint_S \mathbf{ds} \cdot \mathbf{E}_2^s(S) = 4\pi Q.$$

By subtracting both sides of the above equations we get

$$\oint_S \mathbf{dS} \cdot (\mathbf{E}_1^s(S) - \mathbf{E}_2^s(S)) = 0. \tag{A.6.5}$$

Since the above equality holds for an arbitrarily chosen surface inside the considered region, at any point

$$\mathbf{E}_1(\mathbf{r}) = \mathbf{E}_2(\mathbf{r}).$$

This means that the general solution to Poisson's equation (A.6.3) is uniquely specified.

A.7 Discretisation of 2D Laplace's Functional

To represent (14.1.2) after discretisation we replace differential operators with finite differences (using a three-point scheme). Note that not

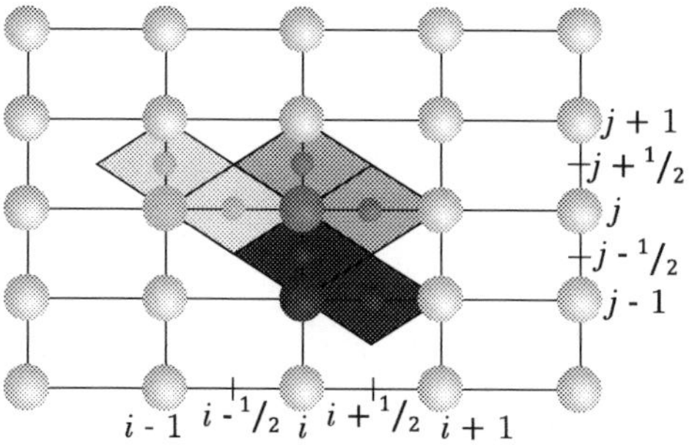

Figure A.4 The grid with marked selected areas

only the grid points are used but also the points halfway between them (Figure A.4).

$$
\left(\frac{\partial \phi}{\partial x}\right)_{i+1/2j} = \frac{\phi_{ij} - \phi_{i-1j}}{h_x},
$$

$$
\left(\frac{\partial \phi}{\partial y}\right)_{ij+1/2} = \frac{\phi_{ij} - \phi_{ij-1}}{h_y}.
$$

This allows us to represent the functional (14.1.2)

$$
F[\phi] = \frac{h_x h_y}{2} \sum_{i=2}^{N_i-1} \sum_{j=2}^{N_j-1} \left[\frac{(\phi_{ij} - \phi_{i-1j})^2}{h_x^2} + \frac{(\phi_{ij} - \phi_{ij-1})^2}{h_y^2} \right]
$$

$$
- h_x h_y \sum_{i=2}^{N_i-1} \sum_{j=2}^{N_j-1} 4\pi \rho_{ij} \phi_{ij}.
$$

The condition for the minimum

$$
\frac{\partial F}{\partial \phi_{ij}} = 0,
$$

leads to

$$
\frac{h_x h_y}{2} \left[\frac{2(\phi_{ij} - \phi_{i-1j})}{h_x^2} + \frac{2(\phi_{ij} - \phi_{ij-1})}{h_y^2} \right.
$$

$$
\left. - \frac{2(\phi_{i+1j} - \phi_{ij})}{h_x^2} - \frac{2(\phi_{ij+1} - \phi_{ij})}{h_y^2} \right] - h_x h_y 4\pi \rho_{ij} = 0,
$$

$$
h_y^2(\phi_{ij} - \phi_{i-1j}) + h_x^2(\phi_{ij} - \phi_{ij-1}) - h_y^2(\phi_{i+1j} - \phi_{ij}) - h_x^2(\phi_{ij+1} - \phi_{ij})
$$
$$
= h_x^2 h_y^2 4\pi \rho_{ij},
$$

$$
2 \left(h_x^2 + h_y^2 \right) \phi_{ij} - \left(h_y^2(\phi_{i-1j} + \phi_{i+1j}) + h_x^2(\phi_{ij-1} + \phi_{ij+1}) \right) = h_x^2 h_y^2 4\pi \rho_{ij}.
$$

This is a starting equation for the Gauss–Seidel iterative minimisation procedure. Let us solve it for ϕ_{ij}

$$\phi_{ij} = \frac{h_x^2 h_y^2 4\pi \rho_{ij} + h_y^2(\phi_{i-1j} + \phi_{i+1j}) + h_x^2(\phi_{ij-1} + \phi_{ij+1})}{2(h_x^2 + h_y^2)}.$$

In the case $h_x = h_y \equiv h$ it becomes simpler

$$\phi_{ij} = \frac{1}{4}\left(h^2 4\pi\rho_{ij} + \phi_{i-1j} + \phi_{i+1j} + \phi_{ij-1} + \phi_{ij+1}\right).$$

A.8 Density of Star

Suppose the model mass density distribution inside a star is given by the formula

$$\rho(r) = \frac{M_\odot}{\pi R_\odot^3}e^{-\frac{2r}{R_\odot}}. \tag{A.8.1}$$

Let us derive the relations used in Project 5. First we calculate mass of the star

$$M = \int \varrho(\mathbf{r})d^3\mathbf{r} = \int \rho(r)4\pi r^2 dr = \int_0^\infty \frac{M_\odot}{\pi R_\odot^3}e^{-\frac{2r}{R_\odot}}4\pi r^2 dr$$

$$= \frac{4M_\odot}{R_\odot^3}\int_0^\infty e^{-\frac{2r}{R_\odot}}r^2 dr = \left| \begin{array}{l} \frac{2r}{R_\odot} = r' \to r = \frac{R_\odot r'}{2} \\ \frac{2}{R_\odot}dr = dr' \to dr = \frac{R_\odot dr'}{2} \end{array} \right|$$

$$= \frac{4M_\odot}{R_\odot^3}\int_0^\infty e^{-r'}\left(\frac{R_\odot r'}{2}\right)^2 \frac{R_\odot dr'}{2} = \frac{M_\odot}{2}\int_0^\infty e^{-r'}(r')^2 dr'$$

$$= \left| \begin{array}{ll} f = (r')^2 & f' = 2r' \\ g = -e^{-r'} & g' = e^{-r'} \end{array} \right|$$

$$= \frac{M_\odot}{2}\left[\left[-(r')^2 e^{-r'}\right]_0^\infty - \int_0^\infty -e^{-r'}2r' dr'\right]$$

$$= M_\odot \int_0^\infty e^{-r'}r' dr' = \left| \begin{array}{ll} f = r' & f' = 1 \\ g = -e^{-r'} & g' = e^{-r'} \end{array} \right|$$

$$= M_\odot\left(\left[-r'e^{-r'}\right]_0^\infty - \int_0^\infty -e^{-r'} dr'\right) = M_\odot \int_0^\infty e^{-r'} dr' = M_\odot,$$

and the contractual radius of the star is

$$\varrho(r) = 4\pi r^2 \rho(r) = \frac{4M_\odot}{R_\odot^3}r^2 e^{-\frac{2r}{R_\odot}},$$

$$\varrho'(r) = \frac{4M_\odot}{R_\odot^3}\left[2re^{-\frac{2r}{R_\odot}} + r^2\left(-\frac{2}{R_\odot}e^{-\frac{2r}{R_\odot}}\right)\right]$$

$$= \frac{8M_\odot}{R_\odot^3} r e^{-\frac{2r}{R_\odot}} \left(1 - \frac{r}{R_\odot}\right),$$

$$\frac{8M_\odot}{R_\odot^3} r e^{-\frac{2R}{R_\odot}} \left(1 - \frac{R}{R_\odot}\right) = 0 \rightarrow R = R_\odot.$$

Poisson's equation is

$$\frac{d^2\varphi(r)}{dr^2} = -4\pi\, Gr \frac{M_\odot}{\pi R_\odot^3} e^{-\frac{2r}{R_\odot}}, \qquad (A.8.2)$$

and the exact solution to this problem is by integration by parts

$$\frac{d\varphi(r)}{dr} = \int -\frac{4GM_\odot}{R_\odot^3} r e^{-\frac{2r}{R_\odot}}\, dr = \left| \begin{array}{l} \frac{2r}{R_\odot} = r' \rightarrow r = \frac{R_\odot r'}{2} \\[4pt] \frac{2}{R_\odot} dr = dr' \rightarrow dr = \frac{R_\odot dr'}{2} \end{array} \right|$$

$$= -\frac{4GM_\odot}{R_\odot^3} \int \frac{R_\odot r'}{2} e^{-r'} \frac{R_\odot dr'}{2} = -\frac{GM_\odot}{R_\odot} \int r' e^{-r'}\, dr'$$

$$= \left| \begin{array}{ll} f = r' & f' = 1 \\ g = -e^{-r'} & g' = e^{-r'} \end{array} \right|$$

$$= -\frac{GM_\odot}{R_\odot} \left[(-r' e^{-r'}) - \int -e^{-r'}\, dr' \right]$$

$$= -\frac{GM_\odot}{R_\odot}(-r' e^{-r'} - e^{-r'}) = \frac{GM_\odot}{R_\odot} e^{-r'}(r' + 1) + C,$$

$$\varphi(r) = \int \frac{GM_\odot}{R_\odot} e^{-r'}(r' + 1) \frac{R_\odot dr'}{2} = \frac{GM_\odot}{2} \int e^{-r'}(r' + 1)\, dr'$$

$$= \left| \begin{array}{ll} f = r' + 1 & f' = 1 \\ g = -e^{-r'} & g' = e^{-r'} \end{array} \right|$$

$$= \frac{GM_\odot}{2} \left[[-(r' + 1)(e^{-r'})] - \int -e^{-r'}\, dr' \right]$$

$$= \frac{GM_\odot}{2}[-(r' + 1)e^{-r'} - e^{-r'}] = -\frac{GM_\odot}{2} e^{-r'}(r' + 2) + Cr' + D$$

$$= -\frac{GM_\odot}{2} e^{-\frac{2r}{R_\odot}} \left(\frac{2r}{R_\odot} + 2 \right) + Cr + D$$

$$= -GM_\odot e^{-\frac{2r}{R_\odot}} \left(\frac{r}{R_\odot} + 1 \right) + Cr + D,$$

from which $\phi(r) = \varphi/r$ follows immediately

$$\phi(r) = -GM_\odot e^{-\frac{2r}{R_\odot}} \left(\frac{1}{R_\odot} + \frac{1}{r} \right) + C + \frac{D}{r}. \qquad (A.8.3)$$

Moreover, we can see that our solution is not unequivocal. Owing to Poisson's equation we should notice that if $g(x)$ is solution, $g(x) + Cx + D$

is solution too ($g''(x) = f(x) \rightarrow (g(x)+Cx+D)'' = g(x)''+(Cx+d)'' = g(x)'' = f(x)$). To determine these constants we use the boundary values potential function.

Following the above solution,

$$\phi(0) = -G \int_\infty^0 \frac{M_\odot}{\pi R_\odot^3} e^{-\frac{2r}{R_\odot}} 4\pi r dr = -\frac{4GM_\odot}{R_\odot^3} \int_\infty^0 e^{-\frac{2r}{R_\odot}} r dr$$

$$= \left| \begin{array}{l} \frac{2r}{R_\odot} = r' \rightarrow r = \frac{R_\odot r'}{2} \\ \frac{2}{R_\odot} dr = dr' \rightarrow dr = \frac{R_\odot dr'}{2} \end{array} \right|$$

$$= -\frac{4GM_\odot}{R_\odot^3} \int_\infty^0 e^{-r'} \frac{R_\odot r'}{2} \frac{R_\odot dr'}{2}$$

$$= -\frac{GM_\odot}{R_\odot} \int_\infty^0 e^{-r'} r' dr' = \left| \begin{array}{ll} f = r' & f' = 1 \\ g = -e^{-r'} & g' = e^{-r'} \end{array} \right|$$

$$= -\frac{GM_\odot}{R_\odot} \left[[-r'e^{-r'}]_\infty^0 - \int_\infty^0 -e^{-r'} dr' \right]$$

$$= \frac{GM_\odot}{R_\odot} e^{-r'} |_\infty^0 = \frac{GM_\odot}{R_\odot}.$$

Thus, for both boundary values

$$\lim_{r\to\infty} \phi(r) = 0$$

$$\lim_{r\to\infty} \left[-GM_\odot e^{-\frac{2r}{R_\odot}} \left(\frac{1}{R_\odot} + \frac{1}{r} \right) + C + \frac{D}{r} \right] = C = 0,$$

$$\lim_{r\to 0} \phi(r) = \frac{GM_\odot}{R_\odot},$$

$$\lim_{r\to 0} \left[-GM_\odot e^{-\frac{2r}{R_\odot}} \left(\frac{1}{R_\odot} + \frac{1}{r} \right) + \frac{D}{r} \right]$$

$$= \lim_{r\to 0} \left[-\frac{GM_\odot}{R_\odot} e^{-\frac{2r}{R_\odot}} + \frac{GM_\odot}{r} \left(\frac{D}{GM_\odot} - e^{-\frac{2r}{R_\odot}} \right) \right]$$

$$= \lim_{r\to 0} \left(-\frac{GM_\odot}{R_\odot} e^{-\frac{2r}{R_\odot}} \right) + \lim_{r\to 0} \left[\frac{GM_\odot}{r} \left(\frac{D}{GM_\odot} - e^{-\frac{2r}{R_\odot}} \right) \right]$$

$$= -\frac{GM_\odot}{R_\odot} + GM_\odot \lim_{r\to 0} \left[\frac{1}{r} \left(\frac{D}{GM_\odot} - e^{-\frac{2r}{R_\odot}} \right) \right].$$

This limit must be finite, thus to apply L'Hopital's rule we must assume

$$\lim_{r\to 0}\left(\frac{D}{GM_\odot} - e^{-\frac{2r}{R_\odot}}\right) = 0,$$

$$\frac{D}{GM_\odot} - 1 = 0 \to D = GM_\odot,$$

then

$$-\frac{GM_\odot}{R_\odot} + GM_\odot \lim_{r\to 0}\left[\frac{1}{r}\left(1 - e^{-\frac{2r}{R_\odot}}\right)\right]$$

$$= -\frac{GM_\odot}{R_\odot} + GM_\odot \lim_{r\to 0}\left(\frac{2}{R_\odot}e^{-\frac{2r}{R_\odot}}\right)$$

$$= -\frac{GM_\odot}{R_\odot} + \frac{2GM_\odot}{R_\odot} = \frac{GM_\odot}{R_\odot}.$$

That shows that both constants fulfil our conditions. In conclusion, we are given exact solution for our problem:

$$\phi(r) = -GM_\odot e^{-\frac{2r}{R_\odot}}\left(\frac{1}{R_\odot} + \frac{1}{r}\right) + \frac{GM_\odot}{R_\odot}\frac{1}{r}, \tag{A.8.4}$$

or

$$\phi(r) = \frac{GM_\odot}{R_\odot}\left[\frac{1}{r}\left(1 - e^{-\frac{2r}{R_\odot}}\right) - e^{-\frac{2r}{R_\odot}}\right]. \tag{A.8.5}$$

When we assume that $G = M_\odot = R_\odot = 1$, then

$$\rho(r) = \frac{1}{\pi}e^{-2r} \to \phi(r) = \frac{1}{r}\left(1 - e^{-2r}\right) - e^{-2r}. \tag{A.8.6}$$

A.9 Energy and Pressure of a Cubic Hydrogen Atom Lattice as a Function of Unit Cell Volume

$V_{cell}(\text{Bohr}^3)$	$E_{total}(\text{Hartree})$	$P(\text{Hartree}/\text{Bohr}^3)$
5.83	0.1350	−0.10400
8.00	−0.0312	−0.05700
10.60	−0.1460	−0.03280
13.80	−0.2260	−0.01960
17.60	−0.2840	−0.01210
22.00	−0.3260	−0.00761
27.00	−0.3570	−0.00490

$V_{cell}(Bohr^3)$	$E_{total}(Hartree)$	$P(Hartree/Bohr^3)$
32.80	−0.3800	−0.00320
39.30	−0.3970	−0.00212
46.70	−0.4100	−0.00142
54.90	−0.4190	−0.00096
64.00	−0.4260	−0.00065
74.10	−0.4320	−0.00044
85.20	−0.4360	−0.00030
97.30	−0.4390	−0.00020
111.00	−0.4410	−0.00014
125.00	−0.4430	−0.00009
141.00	−0.4440	−0.00006
157.00	−0.4450	−0.00004
176.00	−0.4450	−0.00003
195.00	−0.4460	−0.00002
216.00	−0.4460	−0.00001
238.00	−0.4460	0.00000
262.00	−0.4460	0.00000

Further Reading

There are many sources in literature and on the Internet which may serve as additional reading material, to extend and/or deepen the knowledge of the subjects covered in this book. Some examples are given next.

Physics

- **Freedman R. A. and Young H. D.** (2019) *University Physics with Modern Physics*, 15th ed. Harlow: Pearson Education.
- **Griffiths D. J.** (2017) *Introduction to Electrodynamics*, 4th ed. Cambridge: Cambridge University Press. https://doi.org/10.1017/978110 8333511.
- **Fermi E., Pasta J., Ulam S. and Tsingou M.** (1955) *Studies of non Linear Problems.* Document LA-1940, Los Alamos National Laboratory, 978–988. https://doi.org/10.2172/4376203.
- **Dauxois T.** (2008) *Fermi, Pasta, Ulam, and a Mysterious Lady*, Physics Today **61**, 55–57. https://doi.org/10.1063/1.2835154.

Numerical Methods

- **Koonin S. E.** (1998) *Computational Physics: Fortran Version*, 1st ed. CRC Press.
- **Cheney W. and Kincaid D.** (2008) *Numerical Mathematics and Computing*, 6th ed. Belmont: Brooks Cole.
- **Gezerlis A.** (2023) *Numerical Methods in Physics with Python*, 2nd ed. Cambridge: Cambridge University Press.
- **Flannery B. P., Press W. H., Teukolsky S. A. and Vetterling W. T.** (2007) *Numerical Recipes: The Art of Scientific Computing*, 3rd ed. Cambridge: Cambridge University Press.

Index